PRIVATE CITIES

There is hardly another form of urban development that has received so much public attention since the late 1990s as privately organised and secured housing developments. For the antagonists, private communities are icons of post-consensus, fragmenting civic society, enclosing and excluding by contractual constitution and sometimes by walls and gates. For others they are simply an efficient way of organising urban life.

Private Cities brings together an international team of authors in an attempt to construct an interdisciplinary dialogue, based upon empirical evidence, on the global spread of private communities. Case studies from the US, Latin America, the Middle East, Europe and China are used to explore local and global explanations of the phenomenon. The contributions show that private neighbourhoods are a highly significant feature of postmodern urbanism. They are a global phenomenon; they signify a distinct shift in the way cities are organised; they create micro-territories with their own local constitutions; they break down traditional social geographies and create new ones; they provide an alternative model for financing civic goods; they do for home buyers what the package-tour industry has done for individual holiday buyers; they alter patterns of urban insecurities; and they create new micro-societies.

In compiling a range of commentaries from writers taking contrasting perspectives, *Private Cities* challenges readers to reflect on the differences in approach and reasoning.

Georg Glasze is Assistant Professor of Geography at the University of Mainz, Germany; **Chris Webster** is Professor of Urban Planning at Cardiff University, Wales; and **Klaus Frantz** is Professor of Geography at the University of Innsbruck, Austria.

ROUTLEDGE STUDIES IN HUMAN GEOGRAPHY

This series provides a forum for innovative, vibrant and critical debate within Human Geography. Titles will reflect the wealth of research which is taking place in this diverse and ever-expanding field.

Contributions will be drawn from the main sub-disciplines and from innovative areas of work which have no particular sub-disciplinary allegiances.

1. A GEOGRAPHY OF ISLANDS
Small island insularity
Stephen A. Royle

2. CITIZENSHIPS, CONTINGENCY AND THE COUNTRYSIDE
Rights, culture, land and the environment
Gavin Parker

3. THE DIFFERENTIATED COUNTRYSIDE
Jonathan Murdoch, Philip Lowe, Neil Ward and Terry Marsden

4. THE HUMAN GEOGRAPHY OF EAST CENTRAL EUROPE
David Turnock

5. IMAGINED REGIONAL COMMUNITIES
Integration and sovereignty in the global south
James D. Sidaway

6. MAPPING MODERNITIES
Geographies of Central and Eastern Europe 1920–2000
Alan Dingsdale

7. RURAL POVERTY
Marginalisation and exclusion in Britain and the United States
Paul Milbourne

8. POVERTY AND THE THIRD WAY
Colin C. Williams and Jan Windebank

9. AGEING AND PLACE
Edited by Gavin J. Andrews and David R. Phillips

10. GEOGRAPHIES OF COMMODITY CHAINS
Edited by Alex Hughes and Suzanne Reimer

11. QUEERING TOURISM
Paradoxical performances at gay pride parades
Lynda T. Johnston

12. CROSS-CONTINENTAL FOOD CHAINS
Edited by Niels Fold and Bill Pritchard

13. PRIVATE CITIES
Global and local perspectives
Edited by Georg Glasze, Chris Webster and Klaus Frantz

'The most comprehensive survey yet of the rapid global growth of gated neighbourhoods, *Private Cities* presents an important range of case studies which track the privatisation of urban communities across five continents. The book underlines forcefully the diversity and complexity of the gating of communities worldwide.'

Stephen Graham
Durham University

'A balanced, comprehensive, and globally comparative look at the privatization of urban space and the spread of corporate structures into local governance and residential community.'

Edward Soja
University of California, Los Angeles

PRIVATE CITIES

Global and local perspectives

Edited by Georg Glasze, Chris Webster and Klaus Frantz

Routledge
Taylor & Francis Group
LONDON AND NEW YORK

First published 2006
by Routledge
2 Park Square, Milton Park, Abingdon, Oxon OX14 4RN

Simultaneously published in the USA and Canada
by Routledge
711 Third Avenue, New York, NY 10017

Routledge is an imprint of the Taylor & Francis Group, an informa business

First issued in paperback 2011

Typeset in Baskerville by The Running Head Limited, Cambridge

British Library Cataloguing in Publication Data
A catalogue record for this book is available from the British Library

Library of Congress Cataloging in Publication Data
Private cities / edited by Georg Glasze, Chris Webster and Klaus Frantz.
p. cm. — (Routledge studies in human geography)
Includes bibliographical references and index.
1. Gated communities. 2. Gated communities. — Case studies. I. Glasze, Georg.
II. Webster, Chris. III. Frantz, Klaus. IV. Series.
HT169.58.P75 2005
307.77—dc22 2005008571

ISBN13: 9-78-0-415-34170-7 (hbk)

ISBN13: 9-78-0-415-51140-7 (pbk)

CONTENTS

FIGURES AND TABLES

Figures

Tables

CONTRIBUTORS

Sarah Blandy teaches property law at the University of Leeds. Her research interests are the socio-legal aspects of tenure and rights, governance and dispute resolution in housing developments, and the spatial boundaries of law. With Rowland Atkinson, John Flint and Diane Lister, she undertook the first national study of gated communities in England, published by the Office of the Deputy Prime Minister in 2004. She has published widely on gated communities, and was guest editor with Rowland Atkinson of a special issue of *Housing Studies* (2005, vol. 20, no. 2) on this topic.

Axel Borsdorf is a full Professor of Geography, University of Innsbruck, and the director of the Institute for Urban and Regional Research, Austrian Academy of Sciences, Vienna. He has realised studies in geography, geology, early history and German language and literature at Göttingen, Valdivia/Chile and Tübingen. He completed his Ph.D. at Tübingen in 1976. His main research topics are urban, social and cultural geography and mountain research in Europe and Latin America and he is President of the Austrian Geographical Society and member of the board of the Austrian Institute for Latin America.

Jennifer Dixon is Professor of Planning at the University of Auckland. She recently completed a four-year term as Head of the Department of Planning. Her research interests include urban intensification and sustainable design, urban governance and new forms of housing, and strategic environmental assessment. Current projects focus on low-impact urban design, urban form and the management of intensive housing, including gated communities.

Ann Dupuis is Associate Professor of Sociology at Massey University, Auckland, New Zealand. Her research interests in the area of urban studies include housing wealth and inheritance, urban intensification, urban governance issues and gated communities. She has also written on entrepreneurship, labour market change and non-standard work and is currently involved in research on sustainable employment and younger workers.

Fred E. Foldvary teaches economics at Santa Clara University, California. His books include *Soul of Liberty* (Gutenberg Press 1980), *Beyond Neoclassical Economics* (Edward Elgar 1996), *Public Goods and Private Communities* (Edward Elgar 1994), *Dictionary of Free-Market Economics* (Edward Elgar 1998), and (co-edited) *The Half-Life of Policy Rationales* (New York University Press 2003). He also is a senior editor of the online journal www.progress.org. Foldvary's main areas of research include public economics, social ethics, real-estate economics and private communities.

Klaus Frantz teaches human geography at the University of Innsbruck, Austria. His research interest in the field of urban studies is the geography of urbanisation in the United States during the nineteenth and early twentieth centuries and in the last few years his focus has been on gated communities and the privatisation of urban services. He has published widely in international journals and is the author or (co-)editor of seven books.

Guillaume Giroir is Professor in the Department of Geography at the University of Orléans, France. He has published many works – both theoretical and empirical – relating to a geographical approach to transition and globalisation in China. His main areas of interest include Chinese megacities (especially Beijing and Shanghai), satellite towns, state farms, high-tech parks, theme parks and, especially, gated communities (*bieshu qu*). More recently he has studied environmental issues, particularly natural reserves and biodiversity.

Georg Glasze is Assistant Professor at the Institute of Geography, University of Mainz, and teaches human geography at the Institut d'études politiques Paris. He initiated the research network Private Urban Governance/Gated Communities. Glasze is the author of *Die fragmentierte Stadt: Ursachen und Folgen bewachter Wohnkomplexe im Libanon* (Leske and Budrich 2003) and of numerous articles. He was guest editor of theme issues on privately organised, guarded housing complexes in *Environment and Planning B* (with Chris Webster and Klaus Frantz) and in *Geographica Helvetica*. His main research fields are urban studies and social and political geography.

Michael Janoschka is a graduate geographer from Humboldt University, Berlin, with a specialisation in urban and regional planning. He is currently working as a research assistant at the Institute of Cultural Geography at the University of Frankfurt am Main. Recent research fields include insecurity and the sprawl of private neighbourhoods in Latin America, cultural and political impacts of international migration to Spain and shrinking cities in eastern Germany. Janoschka is also a guest lecturer for Masters' studies in urban planning and human geography in Córdoba (Argentina), Alicante and Madrid.

Ulrich Jürgens is Visiting Professor of Geography at the University of Siegen, Germany. He studied geography, economics and political sciences. His Ph.D. was on urban segregation in South Africa, his habilitation thesis on retailing. His main research topics are urban, demographic and economic geography; his main regional interest is southern Africa. He has published lecture books with Jürgen Bähr on *Southern Africa* (Klett-Verlag 2003) and *Regional Urban Geography* (Westermann 2005).

Karina Landman has a background in architecture, urban design and city planning and is currently working for the Programme for Sustainable Human

Settlements in the CSIR Building and Construction Technologies. Areas of specialisation include crime prevention through environmental design (CPTED) and gated communities. She has delivered several papers at national and international conferences, as well as published a number of articles and chapters in international and local journals and books on gated communities and CPTED. Other areas of interest include sustainable development and housing, as well as alternative building technologies.

Renaud Le Goix teaches human and urban geography as an Assistant Professor at the University of Paris 1 Panthéon-Sorbonne. He holds a Ph.D. in geography (University of Paris 1, 2003) and his research about gated communities and private residential governance in the US aims to address the impact of gated communities on segregation patterns. In 2000 he was awarded a Tocqueville Fellowship (French–American Foundation) and in 2002 a Fulbright Research Fellowship (French–American Commission) at the University of California, Los Angeles and at the City University of New York respectively.

Sebastian Lentz is Full Professor for Regional Geography at the University of Leipzig and Director of the Leibniz-Institute for Regional Geography. He has studied geology, geography, German and education at the universities in Heidelberg and Mannheim. His doctoral thesis is on agricultural geography in the Grisons (Switzerland) and his professorial thesis is on development of the private housing market in post-socialist Moscow. Research fields include social geography, urban geography, regional geography, post-socialist transformation, GIS, Europe, successor states of the Soviet Union and high mountain regions.

Setha M. Low is Professor of Environmental Psychology and Anthropology, and Director of the Public Space Research Group at the Graduate Center, City University of New York, where she teaches and trains students in the anthropology of space and place, urban anthropology, culture and environment, and cultural values in historic preservation. Low has been awarded a Getty Fellowship, an NEH fellowship and a Guggenheim fellowship for her work on values and the cultural landscape. She has published widely and lectures internationally on these issues. Her most recent books include *Rethinking Urban Parks: Public Space and Cultural Diversity* (University of Texas Press 2005), *Behind the Gates: Life, Security and the Pursuit of Happiness in Fortress America* (Routledge 2003), *On the Plaza: The Politics of Public Space* (University of Texas Press 2000), *Theorizing the City* (Rutgers University Press 1999), *The Anthropology of Space and Place: Locating Culture* (Blackwell 2003).

Evan McKenzie is Associate Professor of Political Science at the University of Illinois at Chicago and an adjunct faculty member at the John Marshall Law School. He is author of the book *Privatopia: Homeowner Associations and the Rise of Residential Private Government* (Yale University Press 1994) and numerous articles. McKenzie is also an attorney who has represented associations and unit owners in litigation. He is currently studying the evolving relationship between homeowner associations and government, including municipal mandates that require developers to create private governments in new subdivisions, and the trend towards increasing state regulation of residential private governments.

David Parsons works at Sheffield Hallam University where he carries out work on the relationship between property market dynamics and social and economic processes. He has written for a range of journals and books and regularly addresses international conferences. Over the last five years he has been studying the first gated community in Sheffield, investigating the motivations of the occupiers, the financial interests of the developers and the way in which the scheme was assessed according to development policies by the Planning Department.

Rita Raposo is Professor of Sociology at Instituto Superior de Economia e Gestão, Universidade Técnica de Lisboa, Portugal. Her research interests in the area of urban studies include housing markets, urban form and development, ideal images of the space and the city, and gated communities. She also has research interests in the areas of sociology of culture and economic sociology. She has mainly written on gated communities and is currently involved in research on the social construction of trust and gated communities at Centro de Investigação em Sociologia Económica e das Organizações, Universidade Técnica de Lisboa.

Chris Webster is Professor of Urban Planning at Cardiff University; Honorary Professor, School of Real Estate and Construction, Hong Kong University; and Director of the UK Higher Education Academy Centre for Education in the Built Environment. He is co-editor of the journal *Environment and Planning B: Planning and Design* and editor of the journal *Transactions of the Centre for Education in the Built Environment.* He has published over 100 scholarly papers on urban economics and planning and spatial analysis. His latest book, with Lawrence Lai, is *Property Rights, Planning and Markets: Managing Spontaneous Cities* (Edward Elgar 2003).

Rainer Wehrhahn is Professor of Urban and Population Geography at Aachen University (Dept. Geography, RWTH Aachen, Germany). He holds a Ph.D. in geography from the University of Kiel, and has written a post-doctoral thesis about gated communities in Madrid. He has been researcher and visiting professor in South American and southern European universities. His research interests in urban and regional research focus on urban governance issues, gated communities, megacities and migration and integration topics; he is currently involved in research projects on multi-level governance in European cities and on the consequences of demographic change in Germany.

Fulong Wu is a Professor of Chinese Urbanisation and Planning at Cardiff University, UK. He is a prolific writer on urban China with research interests in the evolution of Chinese housing markets, simulation modelling and urban morphology.

Yanjing Zhao is chief planner in the city of Xiamen, Fujian Province, and former director of the China Academy of Urban Planning, and of the Planning History and Theory Research Centre in Beijing.

1

INTRODUCTION

Global and local perspectives on the rise of private neighbourhoods

Georg Glasze, Chris Webster and Klaus Frantz

There is hardly another form of urban development that has received so much public attention since the late 1990s as privately organised, and often secured, housing developments. In the media, in urban social science, as well as in politics and urban planning, there is a lively and controversial debate on the spread of private forms of urban governance. The vigour of the discussion is not surprising given that the enclosure of urban neighbourhoods brings into sharp relief fundamental social questions about private versus public organisation of civic goods and services, the right to a secure environment versus the right to access, communal versus individual consumption, inclusion versus exclusion, heterogeneity versus homogeneity and efficiency versus equity. The debates that have emerged around these issues show that discussions on urban questions are in the end discussions about the society we would like to live in.

The spread of privately governed and secured neighbourhoods (often called *gated communities*) in many countries of the world has frequently been represented as the privatisation of public space and has been associated with a growing local security problem and the importation of commodified neighbourhood values and technology from the USA. However, the empirical bases for these assumptions have for a long time been rather superficial. One aim of this book is to gather together factual information from countries across the globe to present a sketch of the global phenomenon of private neighbourhoods. The picture that emerges is not as simple as may be implied by many commentators. Consider the following, for example:

Private neighbourhoods are a consequence of growing criminality

Growing criminality is often described as the major or single cause of the spread of secured housing estates. Several authors have challenged this interpretation as they suggest that there is not a clear connection between the level of criminality (which is very hard to measure objectively) and feelings of insecurity (for the US and Germany see for example Reuband 1992). Several contributions to this book show that focusing on insecurity as an explanatory factor risks obscuring other factors relating to the demand and supply of urban governance. Security is only one service that residents want and in both conventional and private neighbourhoods it is generally packaged up with other services. Locational choice is made on the basis of subjective evaluation of bundles of civic goods.

Private neighbourhoods involve the privatisation of public space

The value of 'public space' and its endangerment through 'privatisation' is a frequently cited topos within the critique of contemporary urbanism. The reason may be that the concept of 'public space' is a very powerful normative idea (Habermas 1990: 20; Caldeira 1996: 315). However, 'public space' and 'privatisation' are extremely vague analytical categories (Glasze 2001: 161f). It often remains unclear what exactly is privatised; how privatisation is carried out; and what, specifically, is wrong with privatisation. The writings on the privatisation of public space dichotomise the public realm and the private realm and often focus uni-dimensionally on material changes in space. This risks obscuring a more profound and differentiated analysis of the complex socio-economic and socio-political changes associated with private neighbourhoods (Soja 2000: 320). Most so-called public goods in cities (subject to shared consumption) are consumed by 'small publics' and are not truly available to all (Webster 2002: 397f; Webster and Lai 2003: chapters 5–7). Furthermore, the way in which private–public issues are interpreted also depends on whether urban space is literally enclosed by a private neighbourhood (as with road closures). Many master-planned private settlements simply involve the subdivision of a piece of land formerly under single private ownership into many titles under shared ownership. What precisely does this mean for the publicness of the city? From one point of view, the latter process may increase rather than decrease the degree of 'publicness'. A piece of land under single private ownership may become co-owned by many residents.

Private neighbourhoods are a US-American invention

The chapters in this book demonstrate that privately managed, secured and enclosed housing developments are to be found in many regions of the world and that this tendency cannot in all cases be accurately described as a diffusion of the American model. Secured housing developments are not intrinsically an American phenomenon. Neither, more generally, is a territorially fragmented path of urban development distinctly American – an assumption frequently made in French and German discussions that emphasise the integrated nature of European cities. In the rapidly expanding European metropolises of the nineteenth century such as Paris and Berlin, private investors initiated housing schemes that were, in their conceptual design, very similar to many current forms of private neighbourhood (Glasze 2003). London became an inspiration in this, with enclosed and partially guarded residential areas appearing in the early nineteenth century (Atkins 1993). European suburban garden cities were in many senses prototypes for North American gated communities and also for private neighbourhoods in other parts of the world. Digging deeper into local urban morphological history as some of our chapters do, a wide range of regionally specific precursors to contemporary private neighbourhoods is revealed.

Developers of private neighbourhoods are offering a new way of cooperating in cities. Governance is very local; services and facilities are priced by assessments; services are in principle provided in quantities and quality better matched to residents' preferences. For many citizens in many places this is a novel idea. Many are willing to pay the private price. It looks like a major city-changing institutional innovation. The fact that private neighbourhoods have taken off in some places and not in others, however, and the different forms they take and different historical local antecedents they follow indicate the complexity of the phenomenon. The condominium laws governing private governance in Taiwan may look like the French laws that found their way across the Atlantic to the USA in the mid-twentieth century but they reflect particular legal and cultural values and practices in Taiwan (Chen and Webster 2005). And the progression from one set of institutions (local government acts, for example) to another (homeowner governance laws) is a very local affair, explained and shaped by particular path-dependent stories.

What may be said in this respect about the reasons for the rise of gated and other private urban spaces in a country can also be said about their impact on wider urban society. It seems too simplistic to say that they herald the fragmentation of consensual society as we know it and create irreversible territorial social polarisation. The historical examples of enclosed communities documented in a number of papers in this volume did not apparently deal a death blow to urban society. The mutuality of urban living and the functional ties that bind were too powerful to allow that to happen. However, the pessimists on this issue may be right. The costs to society of the increased alienation alleged by some of our contributors to be an inescapable outcome of territorial enclosure may yet prove to be too high. Contractual governments may grind to a halt under the weight of excessive litigation costs – as documented vividly in Evan McKenzie's chapter. Our point is that these processes will play themselves out differently in different social, cultural, economic and institutional contexts. It is the local processes that seem to us to be of greatest interest. Some are undoubtedly influenced by wider global processes. It would be foolish to try and argue, for example, that the high-end condominium sector in world and regional cities does not owe its birth, and to some extent growth, to an internationally mobile global elite labour force.

What we have set out to explore in bringing this collection of chapters together is that private neighbourhoods are emerging in the cities of the world under different sets of influences in different forms and with different effects.

The local manner in which government institutions adapt to and interface with the new institutions of private urban governance will be very important in affecting the balance of effects. Even this is an unpredictable process and therefore worthy of research. Each country has its own story to tell, therefore, and we invite readers to look closely at the evidence presented by each author and to tease out the significant features of each account.

Private Cities brings together 15 contributions from 21 scholars who research private neighbourhoods in different regions of the world. The authors work together in a research network established by the three editors at the First International Workshop on Gated Housing Estates as a Global Phenomenon – held in Hamburg in 1999. Plans for the book were made at a follow-up conference – the International Conference on Private Urban Governance – held in June 2002 at the University of Mainz (Germany).[1] The book is the fruit of a carefully constructed interdisciplinary discussion based upon empirical evidence and frank and open competition between theoretical ideas. The selection of chapters is intended both to record and to extend this debate. Chapters present case studies from around the world and authors use these to explore local and global explanations from a variety of theoretical positions and values. Certain common themes emerge in the authors' discussions and we hope these will challenge readers to look for the commonalities and specificities in each case. We pick up the themes in our Conclusion and offer our own reflections.

A first group of themes relates to the proposition that private neighbourhoods are a manifestation of processes bound up in the globalisation of culture and economy. These include the globalisation of culture and consumer tastes; a global-scale real-estate industry; a general move towards devolution and privatisation of governance functions; globalisation of labour markets, particularly the rise of a wealthy transnational elite; and an increase in social, ethnic, income and life-style heterogeneity within cities – with an attendant rise in fear of 'others'.

A second group of themes relates to the proposition that private neighbourhoods emerge and adapt locally. Themes include the idea that private neighbourhoods are residential clubs, that they emerge to fill a gap left by conventional government, that as a model of sub-local governance they offer advantages over political government in the resolution of conflict, that there are historical and culturally embedded antecedents to modern private-neighbourhood governance, that local manifestations of private communities are shaped by local culture and tastes, that the state institutions that emerge to govern private-neighbourhood markets are also culturally embedded and path-dependent – leading to locally diverse spatial and social outcomes.

The first part of the book contains five chapters that provide new and challenging insights and evaluations on the development of private neighbourhoods in the USA.

Evan McKenzie, lawyer and political scientist, analyses the spreading of common-interest developments (CID) in the USA since the 1970s as a massive privatization of local government functions. He argues that the rise of common-interest housing can be seen as an important part of a larger institutional transformation of urban governance driven by the ideology of privatism – the belief that corporate management is superior to liberal democratic governance. In his view, the gains in terms of economic efficiency that may result from this form of social and governmental organisation must be considered in the light of the apparent costs in the form of increased social, spatial and institutional separation and segregation, and diminished protection for liberal democratic norms of governance.

Economist Fred Foldvary develops quite a different perspective on private neighbourhoods – founded on a belief in the forces of liberal economic organisation. In this perspective private residential governance is distinguished from public-sector or sovereign government in being based on explicit voluntary contracts among persons of equal legal standing. Foldvary sees the growth of private neighbourhoods as an empirical proof of the economic feasibility of private residential governance – the economic basis of private neighbourhoods is the financing of the territorial civic goods (green spaces, security, parking lots and so on) from the rentals generated by these goods. The externalities from public goods that impact a particular territory become capitalised in higher site rentals and site values. For Foldvary, problems such as the growing separation of social groups and high-cost outcomes of restrictive covenants for inhabitants are not immanent to a system of private residential governance but caused by the distorting effects of government regulation and taxation.

For a long time the discussion of private and secured neighbourhoods lacked empirical evidence – even for the often-described case of the USA. The last three chapters in the first part of the book are based on empirical field studies. Setha Low presents the findings of an ethnographic study of residents living in seven urban and suburban gated communities – comparing urban and suburban neighbourhoods in New York City, Texas and Mexico City. Her interviews illustrate three arguments for the rise of private residential governance: social order and moral minimalism in New York City, economic efficiency and tax advantages in Texas and lack of adequate security provided by the state in Mexico City.

The chapter by Klaus Frantz summarises the results of a comprehensive geographical field study on the growth and distribution of gated communities in an American city. Frantz mapped all 641 gated communities in the Phoenix agglomeration and recorded the population of these neighbourhoods. He estimates the population behind gates at 320,000 – 11.7 per cent of the total population of Metro Phoenix. Based on this study, he describes physical and functional characteristics of gated residential developments and comments on where they are to be found, who lives there and why.

Renaud Le Goix discusses the consequences of fading boundaries between public and private management in California when a 'gated community' engages in municipal incorporation. He argues that the sprawl of gated communities is not to be understood as secession from the public authority, but as a public–private partnership: a local game where the 'gated community' provides certain benefits to the public authority, while the property owners' association is granted autonomous local governance. The spillover effects of this structuring of urban space increase segregation, especially when gated communities incorporate, as the municipal institution is instrumental in predating public funds and property for the privilege of a gated enclave.

The second part of the book brings together eight contributions which document the rise of the private neighbourhood in regions where there is already an established debate on this phenomenon (several metropolitan regions in Latin America and South Africa) as well as in regions where the phenomenon has hardly been recorded (Lebanon, China, Russia, Spain and Portugal, New Zealand and England).

Axel Borsdorf and Michael Janoschka present an overview of the sprawl of private guarded neighbourhoods in Latin American city regions since the 1980s. Starting from a historical analysis of urban forms, the current patterns and consequences associated with the spread of gated neighbourhoods are discussed and integrated in a model of the Latin American city structure. Based on narrative and biographical interviews with inhabitants, the two geographers challenge the assumption that the 'condominiumisation' of urban landscapes in Latin America is always caused by crime and insecurity.

In contrast to that conclusion for Latin American cities, Ulrich Jürgens and Karina Landman relate the rise of gated and walled neighbourhoods in South Africa to the increased perception of insecurity and the comparatively high crime rates since the transformation of the apartheid city. The authors discuss reasons for the popularity of so-called walled-in communities as well as the physical, social, political and urban planning consequences for the post-apartheid city.

Georg Glasze discusses the results of a comprehensive field study of guarded residential complexes in Lebanon. He rejects both the interpretation of several commentators who see these complexes as traditional 'oriental' ethnic and confessional segregation in a new guise and universalistic explanations that relate the rise of guarded neighbourhoods vaguely to concepts of globalisation. Glasze develops an approach which analyses the specific social context of urban development as institutions which are locally and historically specific but not immune to wider secular trends. This approach helps to identify the confessionally segmented patterns of social interaction in Lebanon that provide a *sine qua non* context for the rise of guarded residential complexes.

The case of China presents interesting challenges to the discourses on urban enclosure. Guillaume Giroir discusses a highly fortified luxury villa complex in the suburbs of Beijing. He interprets this new kind of housing as a manifestation of the extreme economic disparities that have emerged in China after the economic and political reforms. Complexes such as that studied by Giroir bear clear similarities to the gated communities found in the United States and elsewhere and sit strangely alongside the legacy of the pre-reform Chinese urban landscape.

Chris Webster, Fulong Wu and Yanjing Zhao look beyond China's luxury villa enclaves and develop the thesis that there are many legacy and modern forms of condominium institution in contemporary Chinese society. Co-ownership and co-rule is a ubiquitous institution for organising residential spaces in China and has survived the end of the centrally planned era in a number of intriguing forms. They argue that this makes for a unique set of starting conditions for the future spatial shape of urban China. It also presents a distinctive set of institutional constraints on the evolving internal urban structure of cities.

Sebastian Lentz also presents insights into an economy in transition – Russia. Gated communities first appeared on the Russian property market in the mid-1990s. Since then their numbers have been increasing constantly. Lentz discusses two types of interpretation. According to the first, gated communities are a new phenomenon, related to rapid economic, legal and societal transformation. Prompted by real and imagined threats in a quickly changing, confusing society, they became a successful housing model. The second interpretation points out that guarded housing complexes were already established in Soviet society. In spite of official self-representation to the contrary and widespread perception in the West, Soviet society was deeply fragmented and spatially subdivided. The leading cadres of the Communist Party and the armed forces (*nomenklatura*) had access to elite housing, which included gated communities in the strict sense of the word, from city apartments in big housing blocks to elite dacha colonies.

As we have already mentioned, the guarded and gated complexes in nineteenth-century London may be interpreted as precursors of the boom in the contemporary USA. The recent development of private residential complexes in England and New Zealand is discussed by Sarah Blandy, Jennifer Dixon, Ann Dupuis and David Parsons. Drawing on empirical research in Auckland and in Sheffield the authors first compare the external regulatory environment, including planning law, government policy and property law, that provides the frameworks for the development of private and sometimes gated neighbourhoods in the two countries. Second, they analyse the mechanisms for the internal governance of private residential neighbourhoods. The authors show that the development of private and gated neighbourhoods seems opposed to official government aims and point to the institutional gaps that have allowed these complexes to emerge.

The development of guarded residential complexes in the Iberian Peninsula has sometimes been compared with the boom of gated communities in the US-American Sunbelt states. Rainer Wehrhahn and Rita Raposo present several findings which seem to back this point of view, including the development of second-home complexes in the coastal zones starting in the 1970s and then a spreading of these complexes to central urban areas later on. However, the authors stress that empirical analysis of the recent rise of guarded neighbourhoods in Portugal and Spain reveals a complex set of widely diverging factors to be responsible for the process of privatisation of neighbourhoods in both countries.

Unifying the chapters and straddling the different positions and paradigms adopted by authors is the idea that private neighbourhoods are a new territorial form of political organisation on the local scale. In this sense, we offer the book as a piece of positive urban research. Those who are interested in understanding the causes and consequences will want to reflect on the facts presented and also the diagnosis, explanations and prognosis given in each case. Authors are mixed, in their views of whether private neighbourhoods are good or bad, of where to focus discussion about causality and of the appropriate spatial and temporal scale for evaluating costs and benefits, and in their conclusions – where is it all heading?

Notes

The editors thank Ulrike Mayr (Mainz) for her help as student assistant in editing this book.

1 Abstracts of the contributions to these meetings are available at www.gated-communities.de.

Bibliography

Atkins, P.J. (1993) 'How the West End was won: the struggle to remove street barriers in Victorian London', *Journal of Historical Geography*, 19 (3): 265–77.

Caldeira, T.P.R. (1996) 'Fortified enclaves: the new urban segregation', *Public Culture*, 8: 303–28.

Chen, C.Y. and Webster, C. (2005) 'Homeowners associations, collective action and rent seeking', *Housing Studies*, 20 (2): 205–20.

Glasze, G. (2001) 'Privatisierung öffentlicher Räume? Einkaufszentren, Business Improvement Districts und geschlossene Wohnkomplexe', *Berichte zur deutschen Landeskunde*, 75 (2–3): 160–77.

—— (2003) 'Bewachte Wohnkomplexe und die "europäische Stadt" – eine Einführung', *Geographica Helvetica*, 58 (4): 286–92.

Habermas, J. (1990) 'Vorwort zur Neuauflage', in J. Habermas (ed.) *Strukturwandel der Öffentlichkeit*, Frankfurt: Suhrkamp: 11–50.

Reuband, K.-H. (1992) 'Objektive und subjektive Bedrohung durch Kriminalität. Ein Vergleich der Kriminalitätsfurcht in der Bundesrepublik Deutschland und den USA 1965–1990', *Kölner Zeitschrift für Soziologie und Sozialpsychologie*, 44: 341–53.

Soja, E. (2000) *Postmetropolis: Critical Studies of Cities and Regions*, Oxford and Malden: Blackwell.

Webster, C. (2002) 'Property rights and the public realm: gates, green belts and Gemeinschaft', *Environment and Planning B*, 29 (3): 397–412.

Webster, C. and Lai, L.W.C. (2003) *Property Rights, Planning and Markets: Managing Spontaneous Cities*, Cheltenham and Northampton: Edward Elgar.

2

THE DYNAMICS OF PRIVATOPIA

Private residential governance in the USA

Evan McKenzie

Abstract

This essay presents an overview of private residential governance in the US. It explains what the institution consists of, what forms it takes and how widespread it is; how privately governed communities are brought into existence and how they are regulated. The essay then considers what these new institutions and processes signify in the broader international context within which they must now be placed. Certain incentives and environmental conditions seem to produce and accelerate this transformation of local governance. Yet there are challenges facing these new institutions and questions must be raised concerning the near-term prospects for the success of this experiment with privatising local government. The essay concludes with some thoughts on what the US experience with this institution tells us about the debate over privatisation of public functions.

Introduction

Hardly more than a decade ago the literature on private residential governance was sparse, consisting mainly of government reports, trade association publications, journalistic accounts and some law review articles. Then, beginning in the early 1990s, a number of books appeared that began to create a social scientific context for the subject (Dilger 1992; McKenzie 1994; Barton and Silverman 1994; Blakely and Snyder 1997). Until very recently, nearly all of the available studies dealt mainly with the US. But during the last few years, the research base on private communities has expanded to include an increasing proportion of studies that deal with a wide range of nations on every continent.

We are now able to consider the rise of private urban governance in a comparative perspective and perhaps answer fundamental questions. For example, it is possible that an American institution is being exported to the world, but it may instead be the case that identifiable political, economic and social conditions facilitated the rise of private communities in the US, and are now providing the soil for similar institutions to sprout in other nations. Perhaps both explanations are true to some extent. In any case, the emergence of an international context for the study of private urban governance is a welcome opportunity to reconsider the American experience with private communities.

Privately governed residential communities are now the dominant factor in new housing construction in the US, particularly in the major growth areas of the Sunbelt states and the redevelopment rings of older cities such as Chicago. What were formerly municipal public services and infrastructure are increasingly privatised and placed under the administration of homeowner and condominium associations. All signs point to a continuation and even an acceleration of this trend for the foreseeable future. Many municipalities are now even requiring developers to create associations in all new residential construction, in order to relieve government of the burden of providing services and building infrastructure, while still securing to government the ability to collect a full share of property taxes from the new owners.

These privatised services and infrastructure are thus placed outside the realm of large-scale democratic control and civil liberties, which troubles some observers because it gives rise to the potential for abuse of power and leads to a high level of conflict. On the other hand, it is argued that transferring these functions to the private sphere increases efficiency by subjecting them to market discipline and also increased liberties through freedom of contract and consumer sovereignty (Beito, Gordon and Tabarrok 2002). To their advocates, homeowners' associations (HOAs) are nothing less than a practical Tieboutian solution to the classic economic problem posed by Samuelson of the overproduction of public goods (Samuelson 1954; Tiebout 1956). Some observers envision replacing the municipality as an institutional form with private residential government, not only for new developments, but also for existing neighbourhoods (Nelson 1999). The similarity between private residential governments and libertarian philosopher Robert Nozick's 'private protective associations' (Nozick 1974) is too apparent to go unnoticed.

There is room for a spirited debate over this issue, but it seems to me that the rapid pace at which this transformation has occurred has created a disjunction between practice and policy, and that this gap should concern both the opponents and the supporters of municipal privatisation. The public policy debate over the rise of residential private governance in the US is still in a state of relative infancy, lagging far behind the institutional consensus that spawned the rapid transformation. No consensus has emerged over how to manage the growth or regulate the activities of this new form of local government. Notwithstanding this state of affairs in the US, it appears that many other nations are now heading down the same road and will soon have large numbers of private communities.

In this chapter I will try to present an overview of private residential governance in the US. I will describe how widespread the various forms of private residential governance have become in the US. I will explain what the institution consists of, how privately governed communities are brought into existence and how they are regulated. Lastly, I will consider what these new institutions and processes signify in the broader international context within which they must now be placed. What are the incentives and environmental conditions that produce and accelerate this transformation? What are the challenges and near-term prospects for the success of this experiment with privatising local government? What does the US experience with this institution tell us about the debate over privatisation of public functions?

The scale and scope of private residential government

Several different names for privately governed residential communities appear in the American literature, a fact that exemplifies the still-inchoate nature of this field of study. They are sometimes known as *CIDs*, short for common-interest developments; *community associations*, which is the usual nomenclature used by industry professionals; *HOAs*, short for homeowners' associations; *CICs*, or common-interest communities; *RPGs*, or residential private governments; and in the press and vernacular they are often lumped into the vague category of *gated communities*, even though many do not have gates, walls or other security measures. Common-interest developments is the term that best captures what I see as their essential feature, which is the shared ownership of real property.

Data on the scale and rate of growth of common-interest housing

In the US there are three distinct legal types of common-interest housing: condominiums, housing cooperatives, and planned developments of single family homes run by homeowners' associations. These forms of housing have spread with astonishing rapidity over the US in the last four or five decades and, taken together, these approximately 250,000 developments constitute as of 2003 about 15 per cent of the American residential housing stock, as shown in figure 2.1 overleaf. Of the three types, developments with single family homes and an HOA are by far the most prevalent in new construction, reflecting the overwhelming preference of Americans for detached homes over apartment-like high-density buildings.

Elements of common-interest housing

All three forms of American common-interest housing share certain distinctive characteristics, although there is considerable variation from one development to another as to the precise way these characteristics are expressed. The main elements are common ownership of real property, private land-use controls, private government, master planning and, with increasing prevalence, the use of various security features.

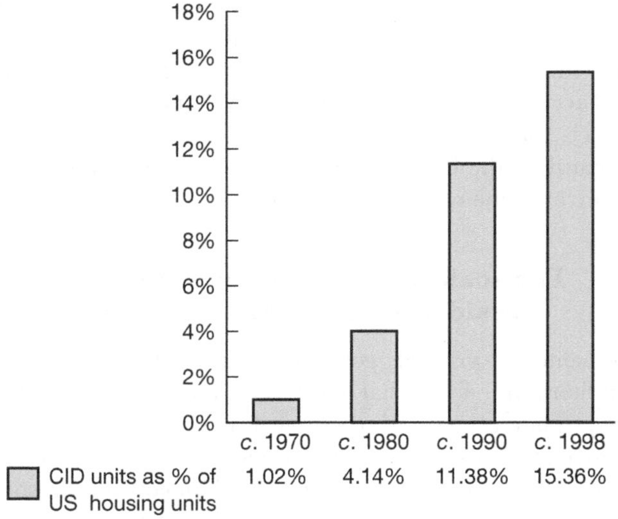

Figure 2.1 The rise of common-interest housing since 1970. Source: Community Associations Institute and United States Census, 1970–2000

Common ownership of real property

What distinguishes common-interest housing most clearly from other housing types is a bifurcated ownership scheme in which all owners have two interests, one an individual interest and the other a common interest. In a *condominium* the individual interest is nothing more than the airspace contained within the owner's unit, along with the inner finished surface of the walls, ceiling and floor. The common interest is the entire remainder of the building or buildings. In other words, the entire condominium development is owned 'in common', in fractional undivided shares, by all the unit owners, with the exception of the airspace within each unit. This notional individual ownership interest is in fact more like an occupancy right. The condominium association does not own any property, but instead manages the building as the representative of all the owners. The condominium is a form of ownership, not a building type, but most condominiums are in fact multi-unit structures that resemble, or once were, apartment buildings.

Housing cooperatives, which are a small fraction of the American market, divide the interest in a slightly different way, although to the eye the buildings typically are indistinguishable from the ordinary condominium. The entire building is owned by the cooperative association, which is typically a non-profit corporation. The owners do not own units, but instead own shares in the corporation. Co-ops are found mainly in high-rise buildings in New York City, with a number in Chicago and a few other older cities.

Developers rarely create them nowadays, preferring instead to use the condominium approach. There are identifiable reasons for the lack of co-ops, and it is a significant issue. Most buyers, and therefore nearly all developers, prefer condominiums over cooperatives because the condominium approach individualises ownership of distinct real property interests, and this facilitates separate financing, insurance and sale. Cooperatives, by contrast, involve buying and selling shares in a corporation that owns the building, and this complicates the legal arrangements for obtaining financing and insurance. Moreover, cooperative associations nearly always have the right to approve or reject potential unit buyers, and this reduces one's individual freedom to sell one's unit. Many buyers find these restrictions burdensome and therefore prefer condominiums, and builders know this.

But there is less conflict in co-ops than in other forms of common interest housing because some sort of meaningful consent can be obtained before a person enters the community. From the standpoint of community harmony this is an advantage over the condominium approach, where there is no way to ensure that owners understand and accept the rules. Yet it also must be noted that the approval requirement in co-ops creates the potential for arbitrary or discriminatory rejection of buyers.

The dearth of co-ops raises questions about the libertarian *voluntary city* explanation for the rise of private communities (Beito, Gordon and Tabarrok 2002). Co-ops are the form of common housing where owner consent is maximised, and where community representatives exercise some control over who joins their community. There is a mutuality of expectations with co-ops that is not found in other forms of common-interest housing. Co-ops are closer in this sense to the assumptions of the voluntary city model. Yet co-ops happen to be the least popular form of common-interest housing, certainly with developers and likely with purchasers as well. The more popular forms – condominiums and planned developments – involve the most minimal and formalistic consent requirements, with membership in the association being automatically created by virtue of ownership alone. The community itself typically has no veto power over the purchase and thus cannot screen out those who do not share or even accept the community's vision or values.

If the rise of common-interest housing were driven by consumer demand for private voluntary cities with a shared lifestyle, one would expect co-ops to be much more prevalent than they are, because co-ops fit that description reasonably well. But if this phenomenon is being driven by the supply-side forces of developer and municipality fiscal concerns, and if homebuyers are for the most part passive consumers of a mass-marketed product, then the prevalence of relatively consent-free common-interest housing makes more sense (McKenzie 1998a, 1998b).

Planned developments of detached single family homes generally include amenities such as private streets, parking lots, and gates and walls; recreation facilities such as swimming pools, lakes, and golf courses; and private infrastructure such as internet systems, water and sewer services, and even private police. These facilities must be owned in common by the residents, with their use managed by some permanent organised body, and their maintenance and repair funded

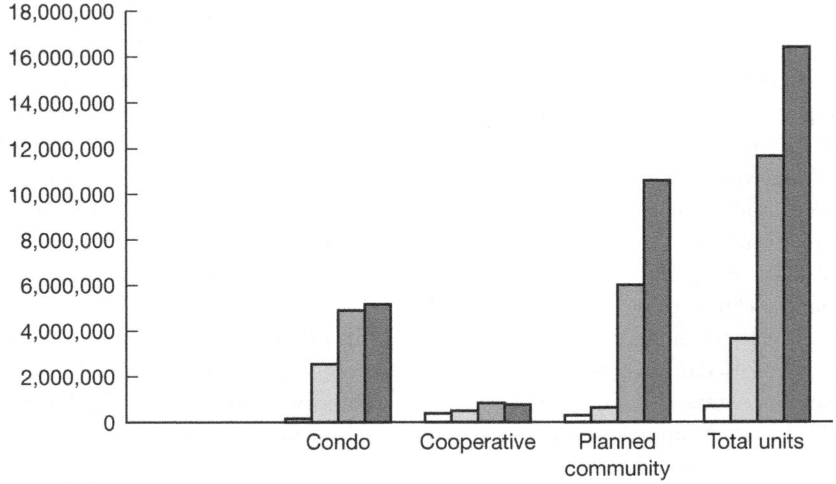

		Condo	Cooperative	Planned community	Total units
1	1970	85,000	351,000	265,000	701,000
2	1980	2,541,000	482,000	613,000	3,636,000
3	1990	4,847,921	824,000	5,967,000	11,638,921
4	1998	5,078,756	748,840	10,562,964	16,390,560

Figure 2.2 Growth of common-interest housing by type, 1970–98. Source: Community Associations Institute

in perpetuity by the owners. This necessitates the creation of a homeowners' association. These associations typically own the common area, unlike the condominium association that merely manages it. For each homeowner, the individual interest consists of his or her house and a lot on which it sits. The common interest is a share membership in the association and thus a voting right in the organisation that owns the common area.

The largest and fastest growing share of common-interest housing is in the planned developments of single family homes, as shown in figure 2.2. This suggests that the rise of common-interest housing in America has changed in a significant way. In the 1960s and 1970s this phenomenon amounted to the spread of new forms of ownership such as condominiums and cooperatives. A new kind of high-density housing was being added to the available stock, but traditional neighbourhoods of single family homes were still the norm. However, from the 1980s on, the rise of common-interest housing has increased to the point where the characteristic American suburban neighbourhood has been replaced by a privatised alternative. The common-interest housing revolution is now more accurately seen as the forced obsolescence of traditional neighbourhoods comprising individually owned suburban homes on public streets with public services. Such places are being gradually – and in some places not so gradually – replaced by privately governed common-interest communities, complete with private streets, services and governments. And in many places this replacement is not just permitted, but mandated, by municipal governments.

Private land-use controls

The means by which these interests are created and the physical properties managed and regulated are contract-based tools for private land-use regulation, generically knows as *governing documents.* Typically there are three main documents for any development.

The first is the articles of incorporation of the condominium, homeowners' or cooperative association. This will set out the powers and duties of the organisation. These are generally quite short, typically just a few pages long. Some associations, typically older ones, are not incorporated, but in recent decades incorporation is practically universal and in some states is required.

The second document, and by the far the most important, is the set of covenants, conditions and restrictions, usually referred to by the acronym CC&Rs. These can be quite voluminous, running into hundreds of pages. Comprising deed restrictions that are also called equitable servitudes, the CC&Rs are the main tool by which developers cast in stone their plan for the project, providing for it to look the same forever.

Deed restrictions are an ancient land-use tool dating back to medieval times, and consist of a specific written limitation on the future use of a piece of property that a seller includes in the deed at the time he or she sells it. The new owner takes the property subject to the limitation on its use, having consented to it by virtue of accepting the deed containing the restriction. For example, if Seller decides that she wishes the property to remain free of commercial uses forever, she places such a restriction on the deed. She has in effect held back that particular right, being unwilling to sell it to any subsequent purchaser. Buyer then acquires the property without having purchased the right to use it commercially. Buyer cannot sell more than he purchased, so the property will never be usable for commercial purposes by any subsequent purchaser.

Multiply this process by several hundred other individual restrictions on use of the property, and you have the means by which CC&Rs operate – owner consent to developer-drafted restrictions. The document will define the common and individual interests, provide for payment of assessments to the association, describe how the association will be run, require maintenance of the common areas and set limits on the use of the individual interests such as describing exactly how they must look and what kinds of modification can be made to them.

In addition, there will typically be a third document called the *bylaws* that will contain details about elections and many other procedural matters. There may also be elaborate sets of architectural restrictions and guidelines, pool or golf course rules, parking regulations and whatever other codifications are deemed necessary to prescribe the limits on conduct within the development.

Private government

One of the most problematic aspects of the rise of common-interest housing is the proliferation of some 250,000 private governments with significant responsibilities and power, staffed by untrained volunteer directors, and operating largely without regulatory oversight. These organisations are exempt from constitutional regulation because they are non-governmental, so that bans on the exercise of what would otherwise be free speech, such as displaying the flag or political signs, are common. They have the power to tax or 'assess' residents to pay for common expenses, on pain of foreclosure and forced sale of the home, and to force them to obey rules by imposing fines. They can file lawsuits in the name of the association, and often sue the developer for defects in original construction, pursue owners for delinquencies in paying assessments or rules violations, and feud with contractors over repair and maintenance work.

In my view, the success or failure of this entire experiment in privatisation will depend on whether these governments can be made to work. To put this in perspective, consider that there are 250,000 association private governments run by about 1.25 million directors and committee members. Their annual operating revenues are estimated at US$30–5 billion, and the aggregate value of the property in CIDs is estimated at US$2 trillion, or some 15 per cent of the total value of all US residential real estate (Community Associations Institute 2004). Virtually all the housing in this sector was constructed since 1970, so there are no traditions or social conventions in place to guide these private governments. Municipalities and state governments offer them little or no support. These volunteer boards have no significant institutional support other than the lawyers and property managers they hire, and these professionals are in turn subject to little governmental oversight. For example, in all but a few states, community association property managers are not even required to be licensed, much less satisfy educational requirements or pass competency examinations. Yet these managers are handling, and sometimes mishandling, billions of dollars of other people's money.

There is no historical parallel of which I am aware for such a rapid and massive transfer of governmental responsibilities, with attendant power and wealth, into the hands of volunteers who are left largely to their own devices in carrying out their appointed tasks.

Master planning

There is nothing new about developers planning large residential neighbourhoods that are specifically designed to meet the needs of an anticipated target population, because it has been done since the early nineteenth century. What is novel about the last few decades of American planned developments is the increasingly precise targeting of a plan and the marketing strategy that supports it to very particular groups of buyers.

A recent study done in California evaluated the degree to which these master-planned developments were promoting segregation or the 'sorting' of people into homogeneous groups by concentrating the 'haves' in particular private communities. The study found that residents of planned developments are, as expected, 'on average white, older, and better educated, and earn higher incomes' (Duncan 2004: 32). The study concludes that these planned developments are contributing only a small percentage of the overall level of segregation in the state. However, the author notes, 'In light of racial and ethnic differences between planned developments and other neighborhoods . . . it is conceivable that as these communities proliferate, they could exert a greater effect on residential segregation over time' (Duncan 2004: 37).

But the level of segregation and its specificity may be too fine-grained for conventional data sources and methods to identify. The practice of 'niche marketing' involves tailoring the physical plan and the CC&Rs to meet the preferences of potential buyers identified by age, race, household size, occupation, current area of residence and place of employment, recreational interests, number and type of vehicles and other factors. The marketing plan then proceeds with an advertising campaign that targets these people by a variety of means, including print advertising, Internet sites, direct mail and road billboards, coupled with the use of public relations to obtain favourable news coverage in the real-estate sections of newspapers and magazines. It is now possible for the supply side of the housing market to produce a remarkably specific segregation of people in ways that self-segregation through individual buyer choice would take ages to generate.

This approach has produced such novel ideas as Front Sight, in Nevada, a development planned for rich gun enthusiasts who seek safety through superior firepower. As noted in the press,

> instead of golf courses and driving ranges, the 550-acre gated community will feature 22 shooting ranges, where residents will be able to pull the triggers of fully automatic Uzis, scale a five-story SWAT tower, practice live-fire drills from a car – and take the kids to school and shop for groceries, all before lunch.
>
> (Wilson 2001)

The plans include a private school for kindergarten through to twelfth grade where, to guard against school shootings by wayward students, the teachers will be armed (op. cit.).

Security measures

During the 1990s it became commonplace for developers to include security features that would previously have been available only to the very affluent, who after all have a long history of living behind walls and gates and protected by armed guards. The striking thing was to see such features become prevalent in middle-class housing (Blakely and Snyder 1997). These accoutrements fall into three categories: *entry controls*, such as gates, traffic barriers, 'no admittance' signs and guardhouses; *hardened perimeters*, including walls, fences and other types of border such as lakes; and *internal surveillance*, which might mean armed patrols but could also be alarm systems or video cameras.

It has been argued that this trend is driven by fears of ethnic diversity at least as much as by fear of crime, and that there may be little or no actual crime reduction involved (Low 2003). Other motivations may be at work as well. It appears that there are many middle-class urban and suburban residents who seek a sense of control over their neighbourhoods and who find security measures and private government the best way of achieving it. Moreover, security features have now become widely viewed as increasing the value of housing, and this contributes to their popularity as well.

The visual impact of such forbidding architectural features has been sufficiently dramatic to give the name 'gated community' to the entire phenomenon of private residential government. However, the variety of approaches to security makes it hard to determine whether developments should really be classified as gated communities, and there is some question over whether the details of such a classification are significant. Some developments have an actual gate, others an occupied or empty guardhouse, and many just have a set of pillars and a 'private property' sign. Do these differences really matter, or is it more significant that the development restricts entry in some fashion, that the streets are private and that there is a neighbourhood private government?

I would argue that it is the creation of private neighbourhoods that deserves our attention more than the existence of gates and walls. Gated communities are more significant as symbols of the phenomenon of which they are a part. These enclosed neighbourhoods, with their fortified borders and 'residents only' signs, serve stark visual notice that present trends seem to be undermining the assumptions of public streets, freedom of movement and access, as well as those of private property and dominion over one's residence that characterised traditional American suburban neighbourhoods.

Creating private governments, American style

Land-use planning in the US is a largely privatised process, but one in which municipalities have considerable power at certain points. One of these points is zoning, when municipalities decide what kinds of development may be permitted in which areas. A second is the setting of specific construction standards through adoption of building codes. Third, and the most significant decision for purposes of a discussion of private urban governance, is the process of deciding whether or not to grant building permits for a developer's plan, and setting conditions that developers must satisfy in order to obtain permission to build. Last is the inspection process, when government employees review the construction as it proceeds to verify compliance with the plans and specifications.

However, the design and planning of residential development, and especially common-interest housing, is still almost entirely the developer's initiative. This pre-eminent role for private planning and capital in the American context often strikes planners and scholars from other nations as a strange way to proceed in that it offers somewhat less opportunity than they would expect for social costs and overall community interests to enter into the calculus over whether and how to proceed. Municipalities have the power to say 'yes' or 'no', within limits, but private developers take the initiative to generate the proposals, raise the capital and carry out the projects.

Within the framework of government prescriptions and approval processes, private real-estate developers then purchase open tracts of land, obtain private financing and offer the specific plans for construction. Municipalities can approve or deny applications to build, and they negotiate with developers for various payments and concessions to cover some of the financial impact the development will have on the community. Once the plans are approved, the developer is in control of the project, subject to inspections by local building authorities for compliance with the plans and local codes.

From a chronological standpoint, this process may begin when a private real-estate development company purchases or options a tract of undeveloped land in a rural or suburban location. Often the seller is disposing of farmland that is no longer profitable to operate as such. Some developers are locals, building a single project or a number in the same area. But increasingly there are a number of large-scale builders in the USA that specialise in the construction of projects that contain hundreds of homes. They have become known as community-builders (Weiss 1987). Firms such as Pulte Homes, KB Homes, the Ryland Group and Hovnanian Enterprises, along with a number of others, are well established in the housing construction industry and have the ability to survive the vagaries of the often turbulent American real-estate market. As one industry insider notes, 'These bigger companies are better able to withstand a regional downturn because of their footprint. Whereas the smaller guys are gonna get clocked' (Robichaux and Pearlman 2004: 69).

Having acquired the land, builders then create a development plan, including both a vision of the physical layout and a clear idea of who will live there and how to market the homes to them. This plan is then presented to the local officials whose approval is necessary, such as a municipal or county planning board or a village council. Then a negotiation process begins in which the local officials may express concerns over the social costs of the project and seek various exactions and concessions from the developer to cover those costs, and often such payments are passed along to the eventual home-buyers as part of the purchase price.

There are legal limits on the nature and extent of these exactions, which must be related and roughly proportional to the actual impact of the project on the community. Unreasonable denial of building permits can be deemed a virtual 'taking' of private property. The Fifth Amendment to the United States Constitution states, 'nor shall private property be taken for public use without just compensation'. Excessive regulation that deprives an owner of virtually all the economic value of the property can be deemed a taking, and developers can sue for damages. In the case of *City of Monterey vs. Del Monte Dunes, 526 US 687* (1999), a residential developer who sought to build several hundred homes on a 37.6-acre plot of beachfront land had a total of 19 different plans rejected five times by municipal authorities mainly on stated grounds of adverse environmental impact. The developer sued and won a jury verdict of US$1.45 million against the city under federal civil rights laws. The United States Supreme Court found this acceptable, holding that developers may use the jury trial process to vindicate their rights in this fashion. Cases such as this lead municipalities to use some caution in the permit review process, because the threat of a large adverse jury verdict is no minor matter.

Developers, of course, do not seek out litigation, and instead have become adept at addressing the arguments against new projects that are often forthcoming from existing residents and their elected representatives. These potential problem areas include environmental concerns; traffic congestion; population pressure on local schools, parks and other facilities; and increased local taxation to accommodate the new residents. Moreover, local governments in many parts of the US are under legal and political constraints that make it hard for them to raise taxes to pay for the costs of growth. This situation became quite severe following the 1978 'property tax revolt' that began in California in which homeowners became highly resistant to increased property taxes (Sears and Citrin 1982).

So developers are able to purchase property from farmers or other existing landowners and local government cannot then prevent them from developing it. Growth will come, and with it will come a host of social costs. Governments find it hard to raise taxes to cover these costs. This situation fuels the proliferation of CIDs in the following manner.

One of the ways developers overcome the fiscal and other objections is to stress that with an HOA-governed development there will be fewer public costs. They argue that much of the new infrastructure that will be needed is to be privately constructed, owned and maintained, and that many of the services that the new residents need will be paid for by their HOA assessments. The residents will also pay their full share of property taxes to local governments.

This argument has become so persuasive that some local governments in Nevada, Arizona and other states have adopted it as their own and are actually requiring that new development be in HOAs. Gilbert, Arizona, located in one of the fastest-growing areas of the nation, for all practical purposes requires homeowner associations in all new subdivisions (Benjamin 2000). In Nevada, both Las Vegas and Henderson do something similar (McKenzie 2005). This situation undermines the voluntarism rationale that legitimises HOA private governance. If only CID housing is available, and if this is largely due to the actions of local governments and large-scale builders, then how can it be argued that the extraordinary limits on owner conduct are justified by consent?

Once the project is approved, the builder obtains financing and assembles a team of professionals to complete the construction plans. This would include architects, engineers, construction contractors and subcontractors for the various trades needed (framers, electricians, plumbers and so forth), as well as the lawyers who will draft the governing documents. Often these documents are standardised, but some developers go to the trouble of hiring law firms that specialise in HOA drafting work and commission an original set of documents that are tailored to the lifestyle foreseen for the particular development and its intended population.

The land is then subdivided into building lots. This is a legal procedure that involves recording the CC&Rs against each of the lots. In form, it is as though a contract had been made between a seller and a host of individual lot purchasers, but in reality the developer owns everything (there are no buyers yet) and the builder is simply taking one parcel and turning it into many, in the process subjecting all the newly created real-estate interests to whatever restrictions he chooses. It is a moment of fictitious agreement that binds all future purchasers of the lots, in perpetuity.

At this point construction can commence, and at the same time typically the sale of lots begins, with model homes used to show buyers what to expect. State laws typically require that a copy of the project's CC&Rs and other governing documents be available in the sales office for potential purchasers to examine. However, there is no way for a buyer to negotiate the terms of these documents because they are already binding on all the lots, and the nature of these interlocking arrangements is such that all lots must be bound by the same terms. Consequently no owner may make an individualised contract to have a different set of covenants than other residents. For example, a 'no pets' restriction cannot be negotiated away for one buyer – it is 'take it or leave it'.

The developer will control the association for a period of time determined by the covenants and by state law. For example, a typical set of covenants might create two classes of voting membership, class A and class B. Class A members are the lot owners who have purchased homes, and they can cast one vote per lot, with no more than one vote per lot allowed no matter how many joint owners of the lot there may be. Class B membership, however, is for the developer-owned units that have not yet been sold, and those will be worth three votes each. So the developer will have majority control over the association for as long as he or she owns at least 17 per cent of the units. This guarantees that the developer will run the association through his or her elected representatives, usually employees, until the development is nearly sold out. The developer's financial interests are protected during this time.

Eventually the day arrives when the developer hands over control of the association to the lot owners. This is a critical moment in the history of any association and is often the start of a time of trouble. The new owner–directors receive the financial records and get a clear picture of the association's economic condition. It is common for developers to keep assessments low in order to make it easier to sell units, and if the new board discovers this they may need to raise assessments. They may also discover defects in original construction that need to be repaired, such as leaky roofs and windows or soil subsidence, and this can lead to conflict with the developer over demands to fix the defects and even to a multi-million-dollar lawsuit. There may also be major policy changes with the new owner-controlled board if the members believe the developer has been too lax or harsh in certain areas. Sometimes these changes lead to conflict between the new directors and owners who are accustomed to the developer's approach to enforcement or assessment. It is also common that the new board will decide to replace any professionals originally retained by the developer, such as a property manager or landscape contractor. These changes can make the turnover period quite turbulent, and for that reason some developers try to prevent that by planning carefully for the transition and bringing owner–directors onto the board during the developer-control period, educating them on the association's affairs, and making the turnover a more gradual process.

What does the American experience tell us?

It is logical to assume that the rise of common-interest housing in the US and elsewhere will continue at the present rate or even accelerate for some time to come, and I do anticipate this, because of the structural forces that favour the trend: rising land costs, decreasing local government fiscal capacity to accommodate growth and consumer preferences for security and neighbourhood control. However, having said that, it is important to at least consider the potential for setbacks or even a reversal of this trend. We cannot be assured that, if the number of CID units almost doubled during the 1990s, it will nearly double yet again in the first decade of the 21st century. This is not a linear process with an inevitable conclusion, but is in fact a deeply and intensely contested process that admits of more than one possible outcome. These issues should be considered by other nations where the American experience is viewed as a model or perhaps a sign of things to come. Established institutions can fail, and there is at least one example of this in the real estate industry in recent memory. Moreover, there are evident problems with the functioning of HOA private governments, and if they are not addressed these challenges could slow or reverse the spread of this form of housing.

Precedent for institutional collapse: the savings and loan implosion

The fact that residential private governance is now an established institution does not mean that it is immune to a sudden reversal of fortunes. Significant and seemingly irreversible trends in human society often encounter circumstances that produce unanticipated sudden change. Sometimes even established institutions fail, and this has happened in the real-estate industry within the last 20 years.

The American savings and loan industry came into existence before the First World War, and was historically focused on lending money to individuals and families for the purchase of single-family homes. It was a federally regulated, low-risk sector of the banking industry that charged low rates on long-term home loans and also paid low interest rates to holders of savings accounts.

Rising interest rates in the late 1970s made these 'thrifts' uncompetitive with other banks. The solution hit upon by the United States Congress was deregulation, including a relaxation of limits on the interest rates thrifts could pay to attract depositors and on the kinds of investment they could make with those funds. During the early 1980s savings and loans were touted as an example of successful rapid deregulation, as evidenced by the rapid expansion of their coffers that followed the lifting of federal restrictions on the interest rates they could pay to depositors.

However, in many cases these federally insured deposits were not invested wisely during this free-wheeling, deregulated heyday, and instead were lost, wasted and stolen. Under the new regime of minimal regulatory oversight, in only a few years the entire industry collapsed in a smouldering heap of financial ruin caused by mismanagement, reckless investment practices and outright criminality. It was 'one of the worst financial disasters of the twentieth century' (Calavita, Pontell and Tillman 1997: 1). A total of 1,043 thrifts failed, worth over US$500 billion. The Federal Savings and Loan Insurance Corporation, which was obligated to pay depositors whose thrift had become insolvent, had nowhere near the assets needed to pay for the mass failure of insured institutions. The cost of the federal bailout that resolved this fiasco was US$153 billion, with 81 per cent of that being borne by the taxpayers and 19 per cent by the industry (Curry and Shibut 2000).

The lesson of this episode should not be lost. A major sector of the American economy that had been a central element of the residential real-estate market for half a century imploded in the course of a few years. It took down with it the federal regulatory apparatus that had been set up decades earlier to guard against such an eventuality. This experiment in deregulation went from apparent success to unmitigated catastrophe in less than a decade.

The savings and loan catastrophe should serve as a cautionary tale for those who insist that simply removing governmental oversight and leaving the market to its own devices will produce beneficial outcomes. It should also offer an object lesson in the rapid changes of course that can affect even large institutional sectors in the real-estate market.

My point is not to predict that HOAs will fail in large numbers, although that is possible, but to point out what issue areas should be addressed to prevent such failures. It is important to stress that we cannot assume today's trends will necessarily extend far into the future if there are systemic problems with the functioning of HOA private governments, and if public policy-makers fail to address these problems.

Specific challenges facing CID housing

At the same time that developers are building record numbers of CID units while municipalities look on approvingly, events are occurring within these developments that raise questions about how smoothly this housing sector is operating. The question marks include a high level of internal conflict, increasingly adversarial press coverage, a trend towards surprisingly intrusive proposals for governmental oversight, and evidence of inadequate planning for major building repairs.

Condo wars

There is considerable conflict between HOA directors and their members that produces a great deal of litigation over parking, pets, children, architectural modifications, assessment levels, alleged violations by directors of the procedures they are supposed to follow, and myriad other details of community life.

These conflicts are inevitable. There will be a good deal of conflict and criticism in, around and about community associations for the foreseeable future no matter what anybody does. This is so for two reasons. First, the nature of the institution requires that neighbours make authoritative decisions about other neighbours' property and home activities, and this generates tension and conflict under even the best of conditions. Second, the increasing market share of this form of housing guarantees that more people will be living in associations who do not understand and accept the rationale on which they are based and who do not have the skills needed to run them.

Americans, to a greater degree than people from most other nations, hold deeply rooted beliefs about the sanctity of private property rights. These beliefs are antithetical to community association life in that people need association approval for doing things, such as painting their home, that would be their sole prerogative in a conventional single-family house. It would be quite a task to replace them with more communal views of property ownership anytime soon. Many, if not most, Americans cannot simply sign away at real estate and close off their actual emotional expectations of what home ownership will feel like. The fact that purchasers have legally consented to a set of enforceable limitations on their use of the property does not suddenly change the emotional reality that they have bought a home and feel like it is really 'theirs', so they may go ahead and put up the holiday lights and the flags, regardless of the rules. Boards of directors will predictably assert their rule enforcement powers in these situations, and from that point on there is a conflict that may end up in court.

A further reason for conflict is the general lack of any set of traditions, institutional constraints or regulatory oversight to guide boards of directors in their rule enforcement activities. There is wide variation in the way these things are done, and any given association can switch from lax or nonexistent enforcement to perceived tyranny, or vice versa, as the result of a single annual election. Furthermore, in the minds of many residents, HOA decisions are not shrouded in the presumption of legitimacy that would promote rapid compliance with a comparable decision by a municipality.

Another reason we can predict high levels of conflict in associations is the increasingly prevalent practice of municipalities mandating that all new residential construction must be privately governed. This practice essentially conscripts all new homeowners in a given area into HOAs, whether or not they wish to be subjected to such potentially intrusive management of their lives and activities. In Gilbert, Arizona questions are being raised concerning the apparent contradiction between the municipality requiring HOAs in new construction, on the one hand, and on the other the municipality having no control over their activities, a situation that, according to one Arizona journalist, 'has defied a logical explanation from the most articulate municipal officials and verbally skilled homeowners association advocates' (Thomason 2004).

The point is that municipalities promote CID construction notwithstanding the fact that there is at present no satisfactory way of resolving member–board conflicts, either within the association or in some other forum, such as alternative dispute resolution, that would be quick and inexpensive and obviate the need to retain legal counsel. The only recourse in most cases is litigation. It would seem that reducing or eliminating the need for costly legal representation is a necessary first step towards a better system of dispute resolution.

In order to reduce the level of conflict in CIDs, the leading trade association in this field, the Community Associations Institute, has recently begun a major campaign to promote a set of guidelines called 'Rights and Responsibilities for Better Communities'. This document is intended to create a consensus and a mutual understanding around the rights and the obligations of both association governing officials and residents (Community Associations Institute 2003). Whether this effort will yield significant success remains to be seen, but this is at best just a good start that by itself could never even approach being a solution. Moreover, this document is an initiative of the national office of the Community Associations Institute, which is unable to compel all those who run the 55 local chapters of that organisation to embrace the document with the same enthusiasm as its authors. The fact is that some of the terms of 'Rights and Responsibilities' are quite unpopular with many local practitioners, perhaps first among those provisions being the statement that foreclosure should be used only as 'a last resort'.

Bad press

Many of these conflicts are reported in the press, to the point where such stories have become a staple of Sunbelt journalism. The press coverage of HOA activities has become increasingly critical, leading to a generally negative public image, especially of the large gated communities. There is a growing public perception that HOA living carries the risk of oppressive over-regulation. There is also concern about the secessionist appearance of a fortified compound for the affluent. Former Secretary of Labor Robert Reich's often-quoted article about the 'secession of the successful' encapsulates this perception (Reich 1991).

Negative press coverage is a significant problem because it can suppress consumer demand, stimulate organised opposition to HOA governance, and lead creative state legislators to think that the time may be opportune to introduce bills aimed at preventing such abuses in the future.

Adversarial legislative oversight?

The political picture of HOA regulation is changing under the pressure of negative press coverage of abuses and also a nascent homeowner rights movement. The Internet has made it possible for small local networks of anti-HOA activists to connect with each other nationally, share information, discuss strategies and advance similar proposals for reform. As a result, the legislatures in several states, including California, Florida, Arizona and Nevada, are being pushed to take a more aggressive and even adversarial approach towards HOA regulation. The most advanced example of this is Nevada, where several cycles of reform over the last ten years have produced an ombudsman's office, a commission, and a number of statutory limits on HOA powers.

It is impossible here to do justice to the variety of issues being considered in a number of other state legislatures. However, some of the proposals were notable for being quite harsh in their potential effects on associations. Bills under consideration in the 2003–4 legislative cycles included an Arizona measure that, in the form that passed the state House of Representatives, would have required HOAs to wait seven years before foreclosing on an owner who did not pay assessments. In Florida legislative proposals included mandatory criminal background checks for HOA directors and criminal prosecution for boards who abuse their powers.

Such legislation could potentially over-regulate CIDS, making the operation of their private governments so complicated that only trained professionals could make them work according to law, and at great expense to owners. It is hard to imagine that this form of private local governance could survive, nor would it make much sense if they could not be run by volunteers. In such a case, there would seem to be little reason not to just use conventional municipalities instead of CIDs.

Follow the money

Many billions of dollars are being spent by untrained HOA directors who have major responsibilities and hardly any oversight, and every dollar comes from homeowners who are often financially strapped, with high mortgage payments and growing property-tax bills.

The financial needs of HOAs increase as the commonly owned property deteriorates with time, a situation that already requires increased assessment levels and large special assessments in many associations. Massive quantities of HOA housing constructed during earlier decades, particularly the boom decade of the 1980s, are just now entering the years when major repairs are certain to be needed.

The majority of associations are not reserving enough funds to cover the cost of major repair and replacement of ageing building components. It is much more common that the assessment levels are geared to covering current expenses and not much more. Given that Americans move about every five to six years, few HOA residents are eager to pay today to replace what will be somebody else's swimming pool ten years from now.

Yet there is no constituency pushing for increased financial prudence that would forestall widespread financial problems. Nobody wants to see assessment levels raised. Developers like to keep assessments low during their control period in order to facilitate sales. Low assessment levels are more attractive to buyers and make it easier for them to qualify for mortgages. HOA owner–directors prefer low assessments because they are paying them as owners, and because other residents pressure them to keep assessments low. Attorneys and property managers may be aware that reserve levels are too low and so advise directors, but even trained professionals are often reluctant to push such unwelcome counsel on their clients for fear of losing their patronage to someone who offers a more sanguine perspective.

Conditions for emergence

This volume tries to address one of the more intriguing puzzles posed by the rapid global spread of private communities, and that is the 'diffusion versus emergence' question. Has an American phenomenon spread across the world, or are there particular economic, political and social conditions in which such communities tend to thrive and which have occurred roughly simultaneously in a number of nations? My own study of the American situation suggests strongly that the latter explanation – emergence in particular environmental conditions – is the more likely explanation. However, diffusion plays a role, in that the existence of an institutionalised (if imperfect) American model, coupled with the ease of exporting the knowledge needed to mass-produce it, makes it easy for developers and planners in other nations to adopt many of the American model's features. It is now unnecessary to invent private local governments. Instead, it is possible to devise local variants on a fully formed institution.

We can conclude that environmental conditions are the most significant variables in explaining the presence or absence of HOA housing in America for two reasons: first, there are wide variations in the amount of such housing across the nation and even within particular states; and second, when we look at places with high and low levels of such housing, there are observable and statistically significant patterns.

The incentives

In my study of California (McKenzie 1998b) I used regression analysis on data from a variety of sources to evaluate the relative contribution of environmental factors to variation in levels of CID construction over a six-year period in the 34 California counties that had any measurable levels of CID housing. I found that

CIDs constituted just under 30 per cent of the new housing sold statewide, with single family homes in planned HOA developments exceeding condominium units at 18.4 per cent versus 11.4 per cent of the new housing stock. Some counties had high levels of CID construction, such as Orange (78 per cent of the new housing), along with Santa Barbara, Los Angeles, Ventura, San Diego, San Mateo, Santa Clara, Alameda, Marin and Contra Costa, all in the 40–50 per cent range. Taken together these counties represent all of the metropolitan coastal counties of the Los Angeles area and the San Francisco Bay area. Outside those two areas, the percentages dropped gradually.

I was able to explain two-thirds of the variation in levels of CID construction with only two variables. One was the relative level of county indebtedness, and the other was the relative cost of urban land. Where counties were heavily indebted, there was more recourse to CIDs in order to minimise the public costs of new development. Where land is more expensive, developers seek out attractive ways to package higher-density housing to maintain profit margins. Where it is less expensive they can still create traditional single-family home developments with larger lots and no association or common areas.

I interpret these findings to mean that local government fiscal constraints and developers' desire to maintain profit levels lead to a public–private convergence that drives the rise of common-interest housing. Consumer preferences and other variables make up the remaining third of the variation.

Privatopia: illiberal dystopia or paradise by contract?

It is always easy for theorists to opine concerning what the world would be like if only they had the power to make their ideas into reality. Rarely in history do circumstances make this happen. But in the rise of common-interest housing we have something like that taking place all around us, not just in America but around the world. A new kind of governing institution that looks much like a private city is spreading rapidly, not because theorists advocated it but because it has suited the short-term needs of developers and local governments. It is not yet possible to evaluate fully the implications of this movement, either in pragmatic terms or as a test of competing theories of governance. We do not yet know whether Nozick and Tiebout were right about the virtues of amplifying the role of individual choice in the provision of public services.

However, some concluding observations are possible. There are some serious departures from the libertarian model in the contemporary American version of CIDs. Perhaps the most significant is the increasing trend towards municipal mandates that force consumers of new housing into CIDs. This seriously undermines the entire voluntarism rationale for the institution. The consequences are that we cannot say how many people really want to live in CIDs; we cannot say they are choosing particular levels of private services for a certain cost, so that the Tieboutian rationale is left in question; and the legitimacy of private rule-making and enforcement is undercut, which is inconsistent with Nozick's voluntary private utopia. This practice of municipal mandates needs to be reconsidered and choice between CID and non-CID new housing restored in those locales that are eliminating such choice.

The problems that are clearly identifiable in American CIDs cohere in a matrix of conflict and litigation, bad press and reactive legislation that bode ill for the future. Unless some way can be found to interrupt this cycle, it is hard to see how the institution can avoid being hampered by annual, piecemeal and over-intrusive reforms. Yet something must be done to protect owner rights against overweening boards and their hired professionals. We must also be careful not to expect too much of the volunteer directors on whom so much depends. The key, it seems, is to find a regulatory model that allows society to test the capacity for efficiency and consumer sovereignty that this experiment in privatisation promises. To date this has not been done in the US, and nations that are considering adoption of US-style models would be well advised to look beyond the surface before concluding that the rapid spread of this form of housing means it must be working well.

Yet the fact of this international spread may prove to be the most positive event in recent years for CID housing in America. It may be necessary to step outside the assumptions of American political culture to see the institution clearly, and arrive at reforms that balance the costs and benefits of individual owners and society.

Bibliography

Barton, G. and Silverman, C. (1994) *Common Interest Communities: Private Governments and the Public Interest*, Berkeley, CA: Institute of Governmental Studies.

Beito, D., Gordon, P. and Tabarrok, A. (2002) *The Voluntary City: Choice, Community, and Civil Society*, Ann Arbor, MI: University of Michigan.

Benjamin, M. (2000) 'Hi, neighbor, want to get together? Let's meet in court!', *U.S. News and World Report*, 129 (17): 56.

Blakely, E. and Snyder, M. (1997) *Fortress America: Gated Communities in the United States*, Washington, DC: Brookings Institution.

Calavita, K., Pontell, H. and Tillman, R. (1997) *Big Money Crime: Fraud and Politics in the Savings and Loan Crisis*, Berkeley, CA: University of California.

Community Associations Institute (2003) *Rights and Responsibilities for Better Communities*, Alexandria, VA: Community Associations Institute. Online. Available HTTP: http://www.caionline.org/rightsandresponsibilities/rights.pdf (accessed 29 August 2004).

—— (2004) *Data on US Community Associations*. Online. Available HTTP: http://www.caionline.org/about/facts.cfm (accessed 29 August 2004).

Curry, T. and Shibut, L. (2000) 'The cost of the savings and loan crisis: Truth and consequences', *FDIC Banking Review*, 13 (2): 26–35. Online. Available HTTP: http://www.fdic.gov/bank/analytical/banking/2000dec/brv13n2_2.pdf (accessed 29 August 2004).

Dilger, R. (1992) *Neighborhood Politics: Residential Community Associations and American Governance*, New York: New York University.

Duncan, T. (2004) *Planned Developments in California: Private Communities and Public Life*, San Francisco, CA: Public Policy Institute of California.

Fishman, R. (1987) *Bourgeois Utopias: The Rise and Fall of Suburbia*, New York: Basic Books.

Jackson, K. (1985) *Crabgrass Frontier: The Suburbanization of the United States*, New York: Oxford University Press.

Jensen, E. (2001) 'Gilbert taking HOAs to task', *Arizona Republic*, 10 May: 1–B.

Low, S.M. (2003) *Behind the Gates: Life, Security and the Pursuit of Happiness in Fortress America*, London: Routledge.

McKenzie, E. (1994) *Privatopia: Homeowner Associations and the Rise of Residential Private Government*, New Haven, CT: Yale University.

—— (1998a) 'Reinventing common interest developments: Reflections on a policy role for the judiciary', *John Marshall Law Review*, 31 (2): 397–427.

—— (1998b) 'Homeowner associations and California politics: An exploratory analysis', *Urban Affairs Review*, 34 (1): 52–75.

—— (2003) 'Common interest housing in the communities of tomorrow', *Housing Policy Debate*, 14 (1–2): 203–34.

—— (2005) 'Constructing the *Pomerium* in Las Vegas: A case study of emerging trends in American gated communities', *Housing Studies*, 20 (2): 187–203.

Nelson, R. (1999) 'Privatizing the neighborhood: A proposal to replace zoning with private collective property rights to existing neighborhoods', *George Mason Law Review*, 7 (4): 827–80.

Nozick, R. (1974) *Anarchy and Utopia*, New York: Basic Books.

Reich, R. (1991) 'Secession of the successful', *New York Times Magazine*, 20 January: 42.

Robichaux, M. and Pearlman, R. (2004) 'How good is your builder?', *Smart Money*, May: 68–9.

Samuelson, P. (1954) 'A pure theory of public expenditure', *Review of Economics and Statistics*, 36 (4): 350–6.

Sears, D. and Citrin, J. (1982) *Tax Revolt: Something for Nothing in California*, Cambridge, MA: Harvard University.

Thomason, A. (2004) 'Crusader does his homework on HOA urban myth', *Arizona Republic*, 12 February: 2.

Tiebout, C. (1956) 'A pure theory of local expenditure', *Journal of Political Economy*, 64: 416–24.

Weiss, M. (1987) *The Rise of the Community Builders: The American Real Estate Industry and Urban Land Planning*, New York: Columbia University Press.

Wilson, M. (2001) 'Home on the range: Nevada firing range owner aims to build luxury desert community for well-to-do, gun-toting set', *San Francisco Chronicle*, 24 June: A–1.

3

THE ECONOMIC CASE FOR PRIVATE RESIDENTIAL GOVERNMENT

Fred E. Foldvary

Abstract

Private governance is distinguished from public-sector or sovereign government in being based on explicit voluntary contracts among persons of equal legal standing. Where homeowners' associations are mandatory or where there is significant government control and taxation of the income or capital goods of civic associations and other proprietary communities, these are to that degree not voluntary and private.

The economic basis of private communities is the financing of the territorial civic goods from the rentals generated by these goods. The externalities from public goods that impact a particular territory become capitalised in higher site rentals and site values. Private governance is efficient and equitable when it taps these generated rentals, in contrast to the typical governmental taxation of transactions and income, which impose an excess burden on the economy.

The economic feasibility of private residential governance, empirically confirmed by the growth of private communities, turns the prevailing 'market failure' doctrine on its head. This doctrine posits that collective goods are subject to free-riders, making the private provision sub-optimal. But the reality of private communities demonstrates the opposite – the governmental provision of public goods such as security and recreation is deficient, and private enterprise has responded to the demand for such by developing private residential communities.

The proposition that 'gated' communities create exclusive zones for the wealthy and fragment the urban community raises the question of cause and effect. To the extent that private communities fill needs and wants not provided by government, the fragmentation is ultimately caused by policy. Gated communities can be the effect, the symptom, of the failure of government to provide adequate security, in addition to the desire of residents to enjoy a greater amount or better quality of club goods than provided by governments.

In the USA, homeowners' associations have been afflicted by conflict over restrictive covenants. The problem cannot simply be ascribed to any inherent flaw in the private-community concept, since private residential communities exist in a setting of government regulation and taxation. We need to analyse whether governmental regulations induce an excessive uniformity of association structures and rules, diminishing free choice; whether the governmental legal system prevents a speedy, affordable and just resolution of disputes; and to what extent unequal taxation, subsidies and legal restrictions prevent a free choice of community formation. Whereas governments save expenses by the private provision of civic services, they usually oppose secession from their fiscal jurisdictions.

Private communities have taken many different forms in the US. These include community residential associations, condominiums, cooperatives, land trusts, apartments, hotels, marinas, mobile-home parks and proprietary communities owned by companies such as Walt Disney World. US residential communities include private neighbourhoods (especially in St Louis), new developments and retirement communities with their own private transit systems. The growth of these diverse private residential communities demonstrates their responsiveness to the desires of the public, and could serve as models for reforming government.

Introduction

Private governance is distinguished from public-sector or sovereign government in being based on explicit voluntary contracts among persons of equal legal standing. Where homeowners' associations are mandatory by law, or where there is significant government control and taxation of the income or capital goods of civic associations and other proprietary communities, these communities are to that degree not voluntary and private.

The economic basis of private communities is the financing of the territorial civic goods from the rentals generated by these goods. The externalities from the public goods that impact a particular territory become capitalised in higher site rentals and site values. Private governance is efficient and equitable when it taps these generated rentals, in contrast to the typical governmental taxation of transactions and income, which imposes an excess burden on the economy, a preventable social cost not offset by any benefits.

The economic feasibility of private residential governance, empirically confirmed by the worldwide growth of private communities, turns the prevailing 'market failure' doctrine on its head. This doctrine posits that collective goods are subject to free-riders, making the private provision sub-optimal. But the reality of private communities demonstrates the opposite. The governmental provision of public goods such as security and recreation is often deficient, and private enterprise has responded to the demand for such by developing private residential communities.

The accusation that 'gated' communities create exclusive zones for the wealthy and fragment the urban community conflates cause and effect. To the extent that private communities fill needs and wants not provided by government, the fragmentation is ultimately caused by government. Gated communities are the effect, the symptom, both of the failure of government to provide adequate security, and of the desire of residents to enjoy a greater amount or better quality of club goods than provided by governments.

In the USA, homeowners' associations have been afflicted by conflict over restrictive covenants as well as litigation against developers for structural flaws. The problem cannot simply be ascribed to any inherent flaw in the private-community concept, since private residential communities exist in a setting of government regulation and taxation. We need to ask whether governmental regulations induce an excessive uniformity of association structures and rules, thus diminishing free choice; whether the governmental legal system prevents a speedy, affordable, and just resolution of disputes; and to what extent unequal taxation and legal restrictions prevent a free choice of community formation.

Market failure versus market success in public goods

The term 'public' has two distinct meanings in economics. With reference to an economy, 'public' means the sector of the economy operated by government. The public sector consists of the enterprises run by government agencies, such as the military, governmental schools and state-run postal services. The rest of the economy is in the 'private' sector. Many governments exert considerable control over nominally private enterprise, but the 'public' sector enterprises are explicitly owned and operated by government officials.

In contrast, 'public goods' are resources, products and services which are used collectively or which impact more than one person. 'Public' in this context refers to people in general, as in 'open to the public'. Public goods can be provided either by government or by private enterprise. It is clearer to refer to such provision as 'collective' in order to avoid confusion between 'public' referring to government (as in 'public schools') and 'public' referring to goods. Private goods are used individually, as when a person eats a sandwich. Again, 'private' in this context is distinct from that of the private sector of the economy, since private goods can be governmental or, like passports, provided by government.

In his seminal paper, Paul Samuelson (1954) presented a mathematical distinction between public and private goods. Private goods are consumed or used individually, so the total amount used equals the sum of the individual usage. In contrast, collective goods are used jointly at the same time, as when a group of people are watching a film. Each person sees the whole film.

Individual consumption or use implies rivalry – the more one person uses, the less is available for others. Collective usage is non-rival – use by one person does not diminish that of another. When the location of the good becomes crowded, then extra users impede use by others, and then some of the characteristics of the item become rival and no longer collective.

Another aspect of goods is exclusivity. An item is excludable if people can be prevented from using it. A park with a fence and gate around it is excludable by physically preventing unpermitted access. All private goods are excludable – if one person uses it, the others have been excluded.

Some economists define public goods as non-excludable. Goods which are jointly used but excludable are put in a third category of 'club goods', a club being an organisation providing excludable goods for its members. But many collective goods not provided by membership clubs are excludable. For example, people can be excluded from highways that are provided by government. Even national defence is excludable, since those outside the country or the defence league are excluded from the defensive service, and people can be excluded from the national territory by being deported. There can also be internal exclusion from services such as streets by putting people in prison. Moreover, exclusion can be a matter of degree, so that if a good such as a neighbourhood street is mostly used by the local residents, others are excluded by the lack of habitual use and the cost of travelling to the locality.

The term 'public' or collective goods will therefore refer here to the Samuelsonian distinction of joint usage, where each person has access to the entire amount, and where the item is non-rival, hence capacious or uncongested.

The meaning and nature of public goods are important because of the widely believed doctrine of 'market failure' for public goods, a doctrine with implications for the services provided by private governance. Samuelson (1958: 334) argues that for public goods, 'we can prove that the perfect-competition market model will not work optimally'. Samuelson (1954: 388) declares also that '*no decentralized price system can serve to determine optimally these levels of collective consumption*' (emphasis in the original). He adds that 'taxing according to a benefit theory of taxation cannot at all solve the computational problem in the decentralized manner possible for the first category of "private" goods'. Textbook writers have generally followed the Samuelson propositions.

The reason for such alleged market failure is the 'free-rider' problem, where people can avoid paying for the good because, once provided, it is non-excludable. Another alleged problem is that there is no way to know how much people actually value the good. With private goods, people will pay for that quantity that they value. With a collective good, there is only one quantity, and there is no way to charge people according to their individual valuations.

There are several premises in market-failure doctrine, usually implicit, which are unwarranted by logic and evidence. These premises of market failure are:

1 Homogeneity

This is the proposition that public goods have a uniform characteristic, namely non-rivalry and non-excludability, that make them treatable as one homogeneous category, rather than having several quite distinct sub-types. If in fact there are significantly different categories of collective goods, then market failure, if it exists, might only apply to some, and not all, categories. Collective goods that impact a particular territory, for example, have different economic effects from those which are non-territorial.

2 Ethereality

This is the implicit proposition that collective goods do not exist in physical time and space. Physicists used to believe in a substance called 'ether', which they thought served as a static medium present throughout the universe, and through which light waves propagate. Physicists now know that ether does not exist. But many economists theorise as though they believe in ether, because that is where they put public goods.

3 Necessarily imposed governance

The market-failure doctrine presumes that market and government are disjoint, whereas the theories of industrial organisation, firms and clubs recognise the obvious fact that there exist private administration and organisational structure in firms and clubs, a private and contractual governance that can provide and manage collective goods.

4 Community atomism

This is the implicit premise that 'the market' consists only of atomistic, unorganised, persons. Hence entrepreneurs who wish to provide collective goods to a large market must contract with a multitude of individuals, and the transaction costs would be too great. But, in historical fact, human beings are social and have always lived in organised communities, and so entrepreneurs need only deal with the governing agents of communities. Moreover, in an interdependent world, the communities themselves are typically federated into higher-level associations, making the entrepreneur's negotiating cost that much less.

5 Homo egotisticus

This is the caricature model of 'economic man', who exclusively and selfishly seeks only a narrow self-gratification, whereas, as Adam Smith (1982, first published 1790) showed long ago in his *Theory of Moral Sentiments*, and as elucidated by Alexander Field in *Altruistically Inclined?*, benevolent sympathy with others is a powerful motivating factor in addition to the self-interest that is more commonly assumed. Indeed, the American economist Henry George (1975, first published 1879) recognised sympathy as potentially more powerful than self-interest.

None of these five market-failure premises have any justification in logic or evidence in the context of civic goods. The key problem in typical market-failure doctrine lies in the implicit ethereal premise. The public goods typically provided by government are located in physical space rather than in ether. Examples include local public goods such as streets and highways, street utilities (lighting) and architecture, parks, security, fire protection, schooling, recreation and public transit such as bus services. Most users are those who are located in the vicinity of the good. Even national defence protects territory, rather than existing abstractly in ether.

To the extent that such collective goods are valued by the residents, they increase the demand to be located near them, relative to not having such goods. This added demand increases the rental that tenants (including owner–tenants) are willing to pay to be located there and the purchase price of land in that territory.

The need to pay rent, either periodically or in the purchase price or as an admission charge, in order to be located in the territory served by the public goods, prevents the resident tenants from being free-riders. Aside from occasional guests, most users are rental-payers. The rental provides the means by which private enterprise can finance the goods. If a firm or club can collect the rental, then they can finance the public good. The market does not fail.

Indeed, the rental is itself a private good, since the totality equals the sum of the individual rentals. The collective aspect of the good is therefore inherently tied to the private aspect of the rental.

The rentals provide the means of determining whether to provide the territorial goods, and the optimal amount. If the rental generated by the goods is greater than the cost, the good is profitably provided. The optimal amount is where the marginal or additional rental generated equals the cost of additional amounts of the goods. For example, an entrepreneur deciding on how many swimming pools to provide to a large community development would, from previous experience and data, provide pools if they add more to site values or rentals than they cost. Since the marginal utility of swimming pools declines with increasing quantity, the profit-maximising developer provides more so long as an extra pool generates more expected rental than it costs.

Territory is inherently excludable, although in some cases the cost of complete exclusion may be higher than the benefits. Territorial clubs include private communities such as condominiums, homeowners' associations, shopping centres, industrial estates or parks, resorts and recreational facilities, office buildings, universities and hotels.

Territory makes club goods excludable in two ways. First, the club can prevent outsiders from entering, such as with a fence and a gate. Second, club services can be de facto excludable if most of the users are paying members, so that free-riders do not materially affect the ability of the club to provide collective goods to the members. A hotel, for example, is financed by the rentals paid by the guests and other tenants, and the fact that non-payers may enter and sit in the lobby or use the elevator does not detract from the financial viability of the hotel.

When government provides the territorial public goods, users are typically double-billed for the cost. Most taxation today falls on wages and capital. A worker–tenant therefore pays taxes on his wages to finance the collective good, and also pays extra rental to a landlord to be located by the territorial goods. If he wants to be located near a metro-rail stop, he pays extra rental relative to locations further away, and he pays sales and wage taxes to the government. To the extent the worker is mobile, a tax on his wage may be passed on to the owner of the firm and to the customers, but typically there are large moving costs and few untaxed options, leaving workers rather immobile for taxes on their wages, and thus forced to bear the burden as a lower net wage.

If some are being double-billed, others must be collecting subsidies on one of those bills. With today's typical government financing, the free-rider is not the user, who has to pay twice for the territorial goods, but the landowners, who get their land values pumped up as the goods become capitalised into higher rentals and purchase prices for real estate. The landlord of a town house near the subway stop receives higher rentals from his tenants due to the public goods paid for by others. This is an implicit redistribution of wealth, a forced transfer of income from worker–tenants to landowners in general, whether they rent out their premises or occupy them as tenant–owners.

When private communities such as homeowners' associations, condominiums or hotels provide territorial goods, they do not enquire about the income earned by the members or tenants, or impose surcharges on the sale of goods in their territory. Hotels instead charge a room rental, and condominiums and associations have a monthly member assessment, which in effect is a site rental payment. They do not have the political power to impose arbitrary costs, so instead they must operate contractually, according to the benefit principle. The rental pays for the territorial benefits.

In private communities, tenant–users therefore pay but once for territorial public goods. They pay only the rental. The providers are thus not free-riders, because in return for the rental payment they receive, they finance the goods, rather than having them provided by some governmental third party. Aside from occasional guests having no material impact, with the private provision of territorial public goods there are neither free-riders nor forced taxpayer drivers.

Why do city, provincial, state and national governments not operate like private communities, financing their territorial goods from site rentals? Such has long been proposed by economists, including the French physiocrats, Adam Smith, Henry George and a number of contemporary public-finance theorists. Why does the public tolerate double billing by government, but does not put up with it in private settings?

Public choice is that branch of economics that studies collective choices, such as voting and the laws enacted by legislatures. The prevailing paradigm of government in the developed world today is mass democracy. A huge mass of voters elect officials whom they do not personally know. The candidates therefore need to send messages to the masses via the mass media, an expensive process that creates a high demand for campaign funds.

Special interests which can receive large benefits, concentrated in their group, have an incentive to organise and provide the funding in exchange for privileges and subsidies, transfers of wealth from the taxpayers and consumers. The government is able to impose tax or restrictive costs on the population at large, with costs spread out thinly so that the typical taxpayer or consumer has little incentive to resist or even obtain the knowledge about such transfers because the benefits of doing so would be mainly external to him.

Do private communities suffer from the same public-choice problem? If they are proprietary, having a single owner such as a corporation, then there is no public-choice problem. For private democratic governance such as in condominiums, the answer is twofold. First, the communities are smaller and their scope more limited, so it is easier to obtain the knowledge of such 'rent-seeking' and to resist. Second, the private communities are not able to impose costs, but must obtain agreement by contract. If a hotel were to request a signed statement of your personal income, the typical customer would feel insulted and take his business elsewhere. But governments routinely do this, and most people submit out of duty or fear. Governments can coerce, while private firms must voluntarily contract.

Of course conflict and exploitation can and do occur in private governance, because the members are somewhat immobile, and some human beings are prone to being opportunist, i.e. prone to exploiting others. Conflicts also occur in households and families, but that does not imply any inherent failure in the institutions of marriage, families or households. But human failure and entrepreneurial failure are distinct from a systematic market failure, where the very structure of an institution or system necessarily makes it fail generally.

In many communities, the opposite of exploitation often takes place. People join voluntary organisations and benevolently contribute their time and resources for the good of the community. The highways of early America, for example, were built privately by civic-minded citizens (Klein 1990). Adam Smith (1982, first published 1790) showed how sympathy with others complements self-interest as a motivator of human action.

The nature of private governance

The economic case for private community governance has a moral foundation, as indeed pure markets do not imply that 'anything goes' but are based on ethical rules. Human action which is peaceful and honest is within the pure free market, while force and fraud are outside the free market and are violations of the ethics of the market.

The British natural-law philosopher John Locke sought to justify government by voluntary agreement. In his *Second Treatise*, Locke (1947: 123, first published 1690) stated,

> The state of nature has a law of nature to govern it which obliges every one; and reason, which is that law, teaches all mankind who will

> but consult it that, being all equal and independent, no one ought to harm another in his life, health, liberty, or possessions . . .

The rules of an organisation have two levels, the constitutional and the operational. A constitution is the supreme law of an organisation, on which all other laws or rules are based. In a homeowners', residential, or civic association, the constitution can be a master deed, which all members explicitly agree to abide by when they sign on as members. The master deed, along with other supreme documents such as by-laws, specify the scope of the organisation, the method of electing representatives of the members to a board of directors, and the powers of the directors. Every organisation, including government, has a constitution, whether it is an explicit document or an implicit understanding.

The operational level of an organisation consists of the budget and the rules the board enacts, by their authority as provided by the constitution. The voluntary nature of a private club is at the constitutional level of agreement. A member agrees to operational rules such as majority voting, where the member may not prefer the particular outcome, but has agreed to the process of achieving the outcome, and thus agrees to abide by the outcome.

The Lockean premises of independence and equality and the ethic they imply are also foundational constitutional agreements for governance. The persons entering a contractual association do so as independent and equal persons, and for the agreement to be ethically valid it must not be imposed on any of the participants. The natural rights derived from natural moral law are inalienable. Therefore a slave contract, permanently impairing such rights, is not ethically valid. This implies that any agreement must allow for the ability of a participant to exit from the association, otherwise it becomes a lifetime slave contract. The agreed constitutional rules can otherwise have any contents that the participants desire. These constitutional principles were stated by Algernon Sidney (1990: 105, first published 1698):

> Those multitudes that enter into such contracts, and thereupon form civil societies, act according to their own will: Those that are engaged in none, take their authority from the law of nature; their rights cannot be limited or diminished by any man, or number of men . . .

Hence contractual communities may be organised according to the voluntary will of the members, so long as inherent rights are respected.

This highlights the key distinction between a private community, such as a residential association, and a government. The difference between a government and a private community is not in the goods provided. The public goods provided by both are similar. The structures of community associations are also similar to those of democratically elected governments.

What places a community organisation in the private or voluntary sector rather than in the government sector is an explicit contract among members with an equal legal status. In contrast, governments impose laws and costs on the members of a community without the explicit consent of all the adult members, and the governing officials are legally superior in having some degree of immunity from lawsuits. The economic theory of clubs presumes such contractual equals. Whereas the conventional market-failure doctrine presumes that markets and governments are distinct, clubs provide market-based governance services to their members in the provision of public goods.

Some scholars have regarded residency in a city or country as implicit agreement with the laws of the realm, including the forms of taxation. There are indeed many implicit aspects to daily commerce, but these exist at the operational level of choice. At the constitutional level, the highest level of choice, for an agreement to be voluntary it must be explicit, preferably on paper or electronic equivalent.

For example, one makes a constitutional choice in getting married. Once married, there are many operational implicit agreements which do not need to be spelled out; indeed, romance may require keeping them unspoken. But getting married should be an explicit choice, not some implicit assumption that then binds the parties into unwanted marriage slavery.

The mere fact that one has not exited from a situation does not imply that one has voluntarily agreed to it. If one falls ill, one could exit the illness by committing suicide. The fact that one has instead chosen to stay alive does not make the illness freely chosen. There would be a truly free choice in governance if people could, without great cost, start new communities free from government control.

In theory, therefore, private communities should create ample public goods, efficiently provided, and with maximum liberty and choice. But in practice, critics have pointed to trouble.

Critics of residential associations have quite correctly pointed to very real and severe problems in today's common-interest developments. The criticism often centres on draconian rules imposed by some boards of such communities, restricting the members' use of their property. Restrictions include seemingly arbitrary covenants and board decisions prohibiting pets, political signs, the flying of flags and even meetings in private homes.

But such problems do not necessarily imply any systematic or necessary failure of private governance. We do not live today in a pure free market, and therefore the outcomes we see cannot be automatically ascribed to the market, or to the concept of private democratic community association. Each problem needs to be analysed to determine whether it is indeed the market, or whether it is interventions into the market, that are the primary cause of the trouble.

Government imposes taxation on labour and enterprise, as well as arbitrary restrictions not based on prohibiting harm to others but on catering to special interests. Zoning restrictions, for example, prevent a wider use of locations for inexpensive mobile homes and, by preventing much development, add to the cost of land. Excessive building codes add to the costs of construction. Taxation imposes large marginal costs and at the same time implicitly grants subsidies to landed interests. Taxation, for example, induced the land-trust village of Arden, Delaware to become municipalised to avoid property taxes on its streets and parks (Foldvary 1994). The American legal system also imposes a huge litigation cost on all enterprise. Taxation, as well as excessively restrictive regulation and excessive litigation of enterprise, skews market prices and profits, creating distorted market signals and thus outcomes that waste resources and thwart people's preferences.

Government is an imposed partner in all property development, not only adding to costs and preventing some options, but also imposing rules on community contracts, making them not truly voluntary. Evan McKenzie, for example, in his book *Privatopia* (1994), noted that the US Federal Housing Administration once actively promoted racial segregation for CIDs.

Much fault also lies with the practice of the real-estate industry in not making explicitly clear to buyers that they are entering into important contractual relationships that will affect their living conditions, much like getting married. Some buyers are rushed and pressured into signing agreements they do not fully understand. A true contract is between fully willing and reasonably informed parties. It verges on fraud to pressure a buyer into membership without his fully knowing the consequences. And fraud is not part of the market. Fraud is theft, and a violation of the market ethic. If there is systematic failure, it is that of government for not providing effective legal remedies for such fraud.

The association in private governance should be a contract among legal equals, but sometimes the members are not truly equals. If boards are given superior powers by the state and some immunity from lawsuits, they become a statist level of government, not a true contractor.

The private community contract, to be truly voluntary, must specify clearly and explicitly the powers of the board, the method of selecting the board, the assessments and fines, the exit options and costs and the method of changing the agreement. All powers of the board must derive only from such constitutional agreement. Too often the board arrogantly usurps power due to vague and overly broad master deed and by-law provisions. Vague and overly broad provisions should not necessarily be prohibited, but their presence should be clearly spelled out so that the buyer knows he will be subject to arbitrary dictatorial operational rule. *Caveat venditor*, responsibility by the seller for a clear explicit agreement, should apply for a truly voluntary acceptance by the prospective member.

Then, too, some mutual associations do not adhere to truly democratic structures. A democratic board should not be self-perpetuating, but elected from any and all of the membership. Requirements to be a candidate should be simple and easy to follow. If such is not the case, the member may reject being a member of a community ruled by oligarchy. If these principles are followed, then it is not the case, as some critics allege, that the private associations violate liberty. Liberty exists at the constitutional level of choice.

Forms of private residential government

Private governance has developed in many forms. The two basic types are proprietary governance, with unified ownership, and democratic governance, with an association of co-owners.

The most common type of proprietary governance consists of real estate owned and operated by a firm, usually a corporation. Spencer MacCallum (2002) calls these 'entrepreneurial communities' or 'entrecoms'. In a land-lease proprietary community, such as a mobile-home park, the company leases lots, on which the tenants construct and own the buildings and other improvements. The company also provides civic services, such as the streets, utilities and security. This is the purest type of land-based financing, where tenants pay a market site rental to occupy a site and use the territorial services.

More commonly, as with multiple-tenant income property, the company rents out the buildings and the land. Examples include company towns, shopping centres and malls, apartment houses, industrial parks, resorts, hotels and marinas. A well-known example is Walt Disney World in Florida (Foldvary 1994). The hotel is the prototype, as analysed by Spencer Heath (1957) and Spencer MacCallum (1970). If the rental amount paid is independent of the income, spending or possessions of the tenant, the rental financing is efficient, because there is no extra charge for personal activities or wealth. In some cases, such as shopping centres, the tenants pay a percentage of their revenue, which may be a way of dealing with the uncertainty of the revenue.

Associations of owners have several forms. In a typical condominium in the USA the association owns nothing. Each member owns his own unit and a share of the common elements; the share of each owner is inscribed in the master deed and is permanent. This share is also the percentage of the vote of the owner for the board of directors. In contrast, in a residential or homeowners' association, the common elements (streets, recreation, transit system, utilities, community buildings) are owned by the association. The members own units such as single houses, and voting for the board can take any form, including possible voting by tenants. There are also civic associations that include business owners, as in, for example, Columbia, Maryland.

A third type of association is a cooperative, where members own shares rather than physical units, and membership entitles one to occupy a unit. Each member has one vote, and new members must be approved by the board, unlike associations and condominiums, where unit sales do not require board approval.

A land trust can be proprietary or it can combine single ownership with a democracy. In the land trust of Arden, Delaware, for example, the trust owns the land and collects the site rentals, but the budget and rules are decided by direct democracy in town meetings (Foldvary 1994). A trust could retain ultimate authority for the use of the site, including redevelopment, while delegating operational authority to the democratic assembly.

As with sovereign government, private governance is subject to exploitation, since moving out is costly. It is therefore important to have safeguards for both the tenants and the owners. Spencer MacCallum (1977) has provided an example of a community governance structure in his constitution for the imaginary community called 'Orbis', which could be a ship, a space colony or a proprietary community on land.

Legal restrictions and tax discrimination have privileged sovereign governance and prevented a greater use of private government. An equal playing field would provide people with the opportunity to choose their governance. Law could facilitate private community formation by enabling residents and enterprises to withdraw from governmental services and replace them with private alternatives, such as private schools and private streets. The members would then be able to deduct the normal community expenses from their tax liabilities. Such secession should be a free choice by all the members in order to be truly private and voluntary, rather than devolving into an agency of the sovereign government. Where only some residents secede, there would be a joint authority for the civic services. The boundaries of the communities would thus be determined by the individual members rather than imposed by fiat.

Wealth and poverty in private residential government

A common criticism of private communities is that they foster a separation of wealthy people in gated communities, apart from the rest of the population, especially the poor. But this too is an outcome of government policy. Almost everywhere in America, for example, private communities have to pay the same taxes to government as non-private communities, even when private communities have their own services, such as streets, which taxpayers then do not have to finance. This makes private communities much less affordable by the poor. If the members could deduct such internal costs from their tax liabilities, and if their assets other than land were tax-free, as are the assets of government, there would be no financial handicap in forming private communities, and the poorer neighbourhoods would then be empowered and able to afford to privatise and communitise their services. The trend could then well be, 'all power to the private neighbourhoods'.

We can foster greater freedom of association with constitutional provisions allowing communities to convert to contractual associations, with a corresponding reduction in tax liabilities. This should be an individual choice; nobody should be forced to join any association, otherwise it would remain in the government sector. If people could choose to associate freely, this would fulfil the dream of social reformers such as Thomas Paine, Ebenezer Howard, Henry George and Spencer MacCallum of a world of consensual communities in levels of association from the neighbourhood to a global association, all freely contracting to provide ample public goods satisfying individual preferences in ways that imposed government can never do.

Bibliography

Field, A. (2001) *Altruistically Inclined? The Behavioral Sciences, Evolutionary Theory, and the Origins of Reciprocity*, Ann Arbor, MI: University of Michigan Press.

Foldvary, F. (1994) *Public Goods and Private Communities*, Aldershot: Edward Elgar Publishing.

George, H. (1975) *Progress and Poverty* (first published 1879), New York: Robert Schalkenbach Foundation.

Heath, S. (1957) *Citadel, Market and Altar*, Baltimore, MD: Science of Society Foundation.

Howard, E. (1965) *Garden Cities of To-Morow* (first published 1902), Cambridge, MA: The MIT Press.

Klein, D. (1990) 'The voluntary provision of public goods? The turnpike companies of early America', *Economic Inquiry*, 28 (4): 788–812.

Locke, J. (1947) *Two Treatises of Government* (first published 1690), New York: Hafner Press.

MacCallum, S. (1970) *The Art of Community*, Menlo Park: Institute for Human Studies.

—— (1977) 'Drafting a constitution for Orbis', *Rampart Individualist*, 1 (1/2): 35–52.

—— (2002) 'The case for land lease versus subdivision: Homeowners' associations reconsidered', in David T. Beito, Peter Gordon and Alexander Tabarrok (eds) *The Voluntary City: Choice, Community, and Civil Society*, Ann Arbor, MI: University of Michigan Press.

McKenzie, E. (1994) *Privatopia: Homeowner Associations and the Rise of Residential Private Government*, New Haven, CT, and London: Yale University Press.

Samuelson, P. (1954) 'The pure theory of public expenditure', *Review of Economics and Statistics*, 36 (4): 387–9.

—— (1958) 'Aspects of public expenditure theories', *Review of Economics and Statistics*, 40 (4): 332–8.

Sidney, A. (1990) *Discourses Concerning Government* (first published 1698), ed. T.G West, Indianapolis: Liberty Classics.

Smith, Adam (1982) *The Theory of Moral Sentiments* (first published 1790), Indianapolis: Liberty Classics.

4

UNLOCKING THE GATED COMMUNITY

Moral minimalism and social (dis)order in gated communities in the United States and Mexico

Setha M. Low

Abstract

This paper reviews theoretical and empirical findings from anthropological and sociological studies that explain the emergence of gating and of private governance. Case studies drawn from an ethnographic study of residents living in seven urban and suburban gated communities illustrate three of the theoretical arguments: social order and moral minimalism in New York City, economic efficiency and tax advantages in Texas, and lack of adequate security provided by the state in Mexico City. The case of the gated community from Mexico City provides a cultural and structural comparison to the US examples since this residential development has created its own system of private governance even though it is illegal to gate public streets and the community administration is not acknowledged by the state.

Introduction

This paper reviews theoretical and empirical findings from anthropological and sociological studies that explain the emergence of gating and of private governance. Case studies drawn from an ethnographic study of residents living in seven urban and suburban gated communities illustrate three of the theoretical arguments: social order and moral minimalism in New York City, economic efficiency and tax advantages in Texas, and lack of security provided by the state in Mexico City. The case of a gated community from Mexico City provides an important comparison to the US examples since this residential development has created its own system of private governance even though it is illegal to gate public streets and the residents' association is not acknowledged by the state.

The gated community is an example of a new form of social ordering called 'spatial governmentality' that focuses on concealing or displacing offensive people or activities rather then eliminating them. Social order is produced by creating zones where the protected group is shielded from others' behaviour. Constance Perin pointed out over 20 years ago that Americans were using land-use controls for symbolic and social ends, excluding people who were not deemed worthy to be included (1977). This new system of socio-spatial regulation promotes safety for the privileged few by excluding those who are considered dangerous, and diminishes the scope of collective responsibility for producing social order characteristic of the modern state.

Urban and suburban spatial separation in the United States has a long history based on racism. Cities continue to experience high levels of residential segregation based on discriminatory real-estate practices and mortgage structures designed to insulate whites from blacks. Since the 1980s there has been a pattern of hyper-segregation in the suburbs, reinforced by patterns of residential mobility by race. Blacks are less likely to move to the suburbs in the first place, and then more likely to return to the city (Denton 1994).

Middle-class and upper middle-class suburban neighbourhoods also exhibit a pattern of class segregation by building fences, cutting off relationships with neighbours and moving out in response to internal problems and conflicts. M.P. Baumgartner identifies this as 'moral minimalism' – suburban residents avoid confrontation and use avoidance rather than settlement or aggression to deal with conflicts (1988). Moral minimalism, however, leads to a decreasing tolerance of heterogeneity, and to the use of external (governmental or institutional) rather than internal (interpersonal) systems of social order. Sally Merry has documented how government has expanded its regulatory role through zoning laws, local police departments, ordinances about dogs, quiet laws and laws against domestic and interpersonal violence that provide a new kind of segregation of family and neighbourhood life (1993). Thus, in the United States class and racial segregation is reinforced by planning, policing and spatial practices, and implemented by zoning laws and regulations. The gated community is an extension of these practices.

The creation of common-interest developments (CIDs) provided a legal framework for the consolidation of residential segregation. A CID is 'a community in which the residents own or control common areas or shared amenities', and which 'carries with it reciprocal rights and obligations enforced by a private governing body' (Judd 1995: 155). Specialised covenants, contracts and deed restrictions that extend forms of collective private land tenure and the notion of private government were adapted by the lawyer and planner Charles Stern Ascher to create the modern institution of the homeowners' association (McKenzie 1994).

Private land-use controls in the United States are not new, but there is a trend away from governmental and public control of land use towards an increased reliance on privately created controls. The shift in the zoning process from a publicly debated and voter-enacted system to a privately imposed system may be far more restrictive than any state statute or local ordinance (Ross *et al.* 1996; Agnotti 1997).

Theories of gating, however, do not necessarily explain the rise of private governance in residential communities in the United States. Robert Nelson points out that in

> 1970 only one per cent of American housing units were located in a homeowner association, condominium or cooperative – the three main instruments of collective private ownership of housing. By 1998, this figure had risen to 15 per cent. In major metropolitan areas, 50 per cent of new housing units are being built and sold as part of a collective housing regime.
>
> (Nelson forthcoming: i)

This increase is a social revolution in governance with private organisations responsible for collecting trash, providing security and maintaining common property. Private enforcement of covenants replaces municipalities regulating the environment by zoning, and new ground rules – voting rights determined by property ownership rather than citizenship – are being put into place. Nelson asserts that the greater flexibility of private governance arrangement has many advantages and should be made available to existing neighbourhoods, replacing zoning controls and privatising zoning. Other theories of private governance include Chris Webster's analyses of the economic efficiency of such forms. He argues it is an efficient solution because of the legal requirement to pay fees, the homogeneity of community needs and desires, and because residents can choose their package of communal goods according to their personal preference (Webster 2001).

What I would like to contribute to this discussion is an ethnographic examination of private governance, specifically homeowners' associations in gated communities, from the point of view of residents living in three suburban areas with contrasting political, legal and socio-economic conditions. Interestingly, residents living in each region have different explanations for why they like having a homeowners' association. Although their narratives do not explain the rapid emergence of the phenomenon, they do contribute to our understanding of its increasing popularity and permanence. Further, the data substantiate a number of the theories of gated community private governance presented above.

Setting

This study began with gaining entry into upper middle- and middle-income gated subdivisions in 1994 and 1995: one in Nassau County on Long Island, and three in the northern suburbs of San Antonio, Texas. Additional gated communities were added: an upper middle- to middle-income community in Mexico City in 1998, and a middle-income and middle-to-lower middle-income community in the New York City area in 2000, to answer questions about class and cultural differences that arose later in the project.

Field sites

I identified new suburban housing developments located approximately 30 to 40 minutes drive from their respective downtown city halls. Each has its own regional style and distinctive design features, but all are enclosed by a five- to six-foot masonry wall or iron fence broken only by entry gates and monitored by a guard or video camera from a central station. All the gated communities studied had additional amenities including swimming pools, club houses, health spas, golf courses and walking trails, and all were landscaped with indigenous trees, grasses, flowers and shrubs maintained by an army of gardeners and automatic sprinkler systems.

Nassau County, Long Island, New York

Nassau County, Long Island has experienced a resurgence of residential development, much of it gated, following the decline in the real-estate market in the early 1990s. With a population of 1,298,842 in 1997, Nassau County abuts the eastern boundary of Queens. Although complete statistics for the number of gated communities in Nassau County have not been compiled, a 2001 survey of the Manor House neighbourhood identified seven gated developments along the main road (Plöger 2002) and in the Pine Hills area, and at least three gated communities located in the immediate vicinity (Low 2003).

In Nassau County gated developments are allowed under incentive zoning regulations recently added to the local village's original zoning and building codes. These changes give considerable economic concessions and freedom in the development of the site. In the case of Manor House, a payment of US$510,000 in compliance with the new zoning code was made to the village to be used to provide public amenities. Most developers in the area have also paid this fee.[1]

Gated communities in the New York area

Manor House

Located in Nassau County, this mostly white and wealthy development of single, detached houses is situated on an old estate with the original manor house retained as a community centre. There is a security guard at the entrance who controls the gate (figure 4.1). The individual houses are large (3,250 to 4,500 square feet), generally two-storey structures, built in a variety of traditional styles, and sell anywhere between US$745,000 and US$1,000,000.[2] The floor plans include three-, four- and five-bedroom homes with a variety of luxury details including a great room with a fireplace and terrace, atrium foyer and large walk-in closet.

Houses are organised along a winding thoroughfare with dead-end streets branching off, leading to groups of houses clustered quite close together on small lots, 15 feet wide, on a quarter to a third of an acre. The remaining

Figure 4.1 Security guardhouse and gate at Manor House. Photo: Jörg Plöger

property near the housing is landscaped to create a park-like atmosphere, and the original site plan left most of the surrounding forested areas unchanged as an amenity for the residents. Since Manor House was developed as a community-interest development, all of the common grounds are maintained by the homeowners' association. The built-out community will contain 141 houses, tennis courts and an outdoor swimming pool, and the mansion has been renovated to accommodate an indoor pool, a billiard saloon, a library, conference rooms, a sauna and cigar-smoking rooms.

Pine Hills

Pine Hills is a gated town-house development of 80 units completed in 1997. There is a gatehouse and residents use electronic identity cards to raise the arm blocking the entrance (figure 4.2). Located near the Long Island Expressway in Nassau County, Pine Hills is a middle-income, mostly white community made up of long rows of similar-looking attached houses in three basic styles. All the houses are two-storey, white and beige clapboard, single-family town houses with wood or stone trim and bay windows (figure 4.3). There are few amenities except for the streets, sidewalks and jogging paths. The location and quality of the town houses have kept the prices in the US$400,000 to US$500,000 range even though it is located in a prestigious, suburban neighbourhood.

Waterview

Waterview is a gated condominium complex of 800 units located in the middle-class neighbourhood of Bayside, Queens. There is both a guarded entrance and

Figure 4.2 Gate at Pine Hills. Photo: Jörg Plöger

Figure 4.3 Town house at Pine Hills. Photo: Andy Kirby

a pedestrian entryway that can be opened with a key. Organised as a series of three-storey buildings, each with three apartments arranged around a swimming pool with health club facilities and sauna, Waterview has been successful in maintaining its prices in the US$350,000 to US$450,000 range. Many of the apartments are rented by their owners, so that there is a broad spectrum of residents of various income levels and greater cultural diversity including African nationals, African-Americans, Korean-Americans and white workers and professionals.

The apartments are small (1,200 square feet) in comparison to the houses in the other communities, although some residents purchase all three levels and reconfigure the space as a single town house. The normal configuration, however, is a one-bedroom apartment with a small garden enclosure and garage on the first floor, a two-bedroom apartment with a bay window and rights to park on the street on the second floor, and a third-floor apartment with two bedrooms and a balcony with rights to park on the cement driveway in front the house.

San Antonio, Texas

San Antonio is divided by Loop 410, a highway that circles the city and defines differences in quality of life, schools, housing developments and municipal services. Inside the loop, 96 per cent are non-Anglo and poor, and house prices average US$20,000, while beyond Loop 410 is mostly Anglo and wealthy, and homes in newly constructed gated communities average US$200,000. Inside Loop 410, 43.5 per cent of people mention crime as a neighbourhood problem, contrasted with 26 per cent of residents outside the loop. The area north of Loop 410, originally five ranches in 1983, is now one of the fastest-growing residential areas in the country (MacCormack 2000).

Like Nassau County, many of the incorporated towns north of Loop 410 have amended their zoning ordinance to allow private subdivisions, but erecting gates around neighbourhoods can be costly for residents, because they must pay for street maintenance and other expenses. Even though expensive, the demand for gating is increasing (Graham 1999).

Gated communities in the San Antonio area

Sun Meadow

Sun Meadow is part of the master-planned suburban development centred on a private golf and tennis club with swimming pools, restaurant and clubhouse. The subdivision includes 120 lots, a few fronting the golf course, surrounded by a six-foot masonry wall. The main entrance is controlled by a grid design gate that is opened electronically by a hand transmitter or by a guard who is contacted by an intercom and video camera connection (figure 4.4 overleaf).

The broad entrance road divides into two sections leading to a series of short streets ending in cul-de-sacs. The single family detached houses are large (3,000–6,000 square feet) two-storey brick colonials or stucco Scottsdale designs with a few one-storey brick ranch-style houses, and sell for between US$275,000 and US$650,000 (figure 4.5). A wide variety of house plans are offered from the three different homebuilders including a two-storey, four-bedroom family model with the master bedroom on the first floor, a two-storey model with a game room, and a single-level four-bedroom house with a 'morning' room, family room and 'owner's retreat'. These different models accommodate the mostly white and a few Hispanic upper middle- and middle-income families. A small number of upper-class Mexican nationals also live there part of the year.

The Links

The Links is a mostly white and upper-income luxury gated community with a 24-hour guarded entrance and a divided main thoroughfare filled with ponds and lush vegetation. All of the 22 lots face the golf course. The single family houses are large (4,000–6,000 square feet) designed by individual architects to look like small mansions with French, Italian or Southwestern architectural details, similar to those found in wealthy areas of Beverly Hills or Palm Springs, California and other upper-income resorts. They sell for between US$500,000 and US$750,000.

The Lakes

The Lakes is a white and Hispanic middle-income gated community in a north-western central suburb of San Antonio. The 59 lots are organised along a series of curving roads with three prized cul-de-sac neighbourhoods. Located on a hill that looks out over the city, these lots are smaller and include three blocks of attached town houses among the single family detached homes. The mixture of house types is unusual, and lower the overall house prices to the US$250,000 to US$300,000 level. Three house styles are available: brick colonial, modern ranch, and Southwestern.

Mexico City

Neighbourhoods are organised along class lines with the upper classes located along a vertical spine to the north and the south of the city. The outskirts of the city, previously inhabited by the illegal squatter settlements of the lower class and poor, are being expropriated, and public land has been privatised, with installed services, and sold to individuals. There are a number of gated communities in these growing residential areas.[3]

Figure 4.4 Cattle gate at Sun Meadow. Photo: Joel Lefkowitz

Figure 4.5 Scottsdale-style house at Sun Meadow. Photo: Setha Low

Vista Mar

Located on the southern edge of Mexico City at the base of Ajusco Mountain, off the major urban highway, the Periférico, Vista Mar was planned to be a part of the Tlalpan Forest, one of the most important ecological reserves in the city. This upper-middle-class community has four access gates controlled by armed guards and is surrounded by forest and a stone wall. No two roads cross, in order to prevent accidents, and all streets end in a T to discourage automobiles driving by. Roads meander through the hilly terrain dotted with parks and gardens landscaped with semi-arid plants and the local volcanic rocks.

House size varies considerably as well as styles that include classic modern with straight lines and volcanic rock details, Mexican colonial, and modern/postmodern designs that use contrasting shapes in the façade. A typical house has two storeys, with three bedrooms on the upper floor each with a separate bathroom, and on the lower floor a studio or library, television room, as well as a living room, dining room, kitchen, breakfast area and living spaces for domestic employees.

The community is divided into 900 lots, 400 square metres each, and families purchase up to three lots for an individual house. A 400-square-metre lot costs 2,000,000 pesos (US$120,000), and houses on single lots have been sold for between 3,500,000 and 4,000,000 pesos (US$420,000 to US$500,000). This gated community does include a few businesses – about 5 per cent of the community – including a bank and a hotel, and a few apartments owned by the tenants.

Findings from the comparative gated community study

Six of these seven communities are located in the suburbs or at the edge of the central city. In San Antonio, the gated communities studied have been recently annexed and are now part of San Antonio, while in New York, except for Waterview in Queens, the developments are located in townships within Nassau County. In Mexico City, as in San Antonio, the periphery is being incorporated in the city, although these neighbourhoods retain the density and ambience of suburban environments. Gated communities also share a number of suburban demographic characteristics including lower population density, lower crime rates, higher percentages of middle-class and upper middle-class professionals, and better schools and services, and are disproportionately white or European when compared to adjoining urban areas. Many of the reasons people and businesses leave the city for the suburbs – to avoid negative factors such as crime and high energy costs, to benefit from lower tax burdens, to have a better standard of living and a wider range of amenities and to profit from local governmental policies – are similar to those people give for choosing gating. In myriad ways, then, gated communities are a special kind of suburb with many of the same features and benefits as those found in non-gated, suburban areas (Muller 1996).

Moral codes and social order: Nassau County, Long Island, New York

Constance Perin argues that all land-use practices – planning, zoning, and development – act as moral codes for unstated rules governing what are widely regarded as correct social categories and relationships. By examining these codes she uncovers Americans ideas of how society is and ought to be organised (Perin 1977). She illustrates how land-use practices are sustained by the economic impact of the social meanings of home ownership. For example, along the expanding suburban frontier the distinction between newcomers and old-timers, as well as between those people who stay and those who leave, dominates. Within this ostensibly moral scheme, newcomers and transients are suspect because of their lack of commitment to the neighbourhood, their instability and other associated characterisations. Newcomers and transients can also lower property values and the quality of life and thus pose an economic as well as a social threat. Regulations in the form of zoning codes and the covenants and restrictions of homeowners' associations work to legitimise, even 'sanctify', homeowners' rights to a protected investment. Couched in moral terms are very real economic concerns (Perin 1977).

M.P. Baumgartner also sees upper middle- and middle-class suburbs in the United States as moral systems, organised to promote what she calls 'moral minimalism' controlling conflict within the community only by external and structural means. She contends that suburbs are physically and socially structured to provide privacy and separation, and households maintain considerable social distance. Neighbours have what she defines as 'weak ties' to other people, but although not strongly linked they can recognise one another. They monitor their environments closely, 'identifying those who do not belong', but are sheltered by the privacy made possible by this freedom of association (Baumgartner 1988: 10).

Thus suburbs are a distinctive kind of neighbourhood that combines diffuse interpersonal associations, homogeneity, autonomy and independence with an absence of strangers. This pattern of social morphology allows suburbanites to keep conflict at a minimum, and to manage problems that do arise by avoidance rather than confrontation. Yet 'moral minimalism' results in a distaste for the pursuit of grievances or censure of wrongdoing, and 'only when they can be assured that someone else will bear the full burden of moral authority, allowing them to remain completely anonymous and uninvolved, do suburbanites approve the exercise of social control' (Baumgartner 1988: 127).

Private governance enhances the ability of suburbanites to keep interpersonal and neighbourhood conflict at a minimum. The complex rules and regulations guarantee that most problems are resolved before they start. In the New York area both the homeowners' associations and the village and municipal organisations work together to reduce intra-community conflict, and the covenants extend and elaborate land-use zoning codes to incorporate an ever widening set of moral and social concerns.

Moral minimalism and the use of land-use planning and zoning codes as moral codes are reflected both at the level of local governance and in the preferences and concerns of gated-community residents. For example, based on his research on gated communities in the village of North Hills, Long Island, Jörg Plöger (2002) found disputes between residents, supposed to be resolved by the homeowners' association, that end up in the village court. The reason this happens is that 'it is often hard to say who's responsible' (fieldnotes of Jörg Plöger). For example, Manor House had a mosquito problem, which occurred when one homeowner, whose garden is situated a few feet higher up, watered her plants too much. The water kept accumulating in puddles in her neighbours' lot which became a breeding pond for mosquitoes for which the lower neighbour was blamed. The neighbours could not resolve the problem so the managing agent had to call up the village court to make the decision. Yet while the village has to resolve intra-community conflicts, it has also given up almost all of its municipal duties, so the homeowners' association remains powerful in other ways.

Incentive zoning practices also reflect the ongoing negotiation of an unstated moral code. Developers usually approach the village board saying, 'We need more flexibility in designing our subdivisions' (fieldnotes of Jörg Plöger). The village is willing to allow some relief, but wants to receive some benefit in exchange. Legally the village cannot negotiate the fees paid in lieu of creating a public amenity. But in practice negotiations with the village mayor occur throughout the review process, and in the end a compromise is reached that satisfies both parties. Thus the fees paid to the village for zoning changes are completely negotiable, and there are no standards for the calculation of the fee. Every single item on the building application list, including the fencing, has to be approved or negotiated against a fee. This exchange is considered acceptable because North Hills has very low taxes compared to other municipalities and keeps the tax burden to a minimum. These low taxes are made possible by offering only limited public services, and they use the fees paid by developers in exchange for incentive zoning to create public amenities. The reason developers who were allowed incentive zoning always paid a fee and never actually created any public amenity is that the village of North Hills has little public land and is almost entirely privately owned.

New York gated-community residents also have moral concerns that are played out in terms of an aesthetic code that conceals their social and economic objectives, rather than moral minimalism. For example, Kerry is pleased with the controls exerted by her homeowners' association and its regulations. She is a gardener and was delighted there would be some areas for planting flowers at her Manor House home. When I notice all the flowers, she explains,

Kerry: They're [the homeowners' association maintenance company] putting in a garden of shrubs and trees in the back, but we're going to do all the flowers, and they will do the twice-a-year maintenance. So there is a certain amount of leeway that they allow us on our property. I mean there are certain things you can not do. 'Well gee, can I put up a little gazebo in our backyard?' [Kerry mimics herself.] 'No.' [Kerry responds as if she is the homeowners' association] There are certain structures you can not build. Which I understood going in [to the community]. There are limitations. The fact that people can't do whatever they want makes me happy, because I don't want ugly structures. Someone might think it's beautiful, and I think it's atrocious. And I'll have to look at it.

Setha: You don't mind as long as you know that other people are not going to put up something as well.

Kerry: Exactly. If it goes for them, it goes for me too. So, that's fine.

Iris and Rebecca agree. As part of their desire to live in what they consider a perfect community and to retain its 'niceness', they tolerate the restrictions of homeowners' associations. Iris comments that a homeowners' association 'is a pain in the neck', but it protects you from the neighbours who want to paint their house red. She chose Manor House for the gating, as well as for the homeowners' association rules and regulations so that the community would retain its original style. A desirable gated community for these families includes control of the built environment, and keeping the landscape, architecture and decorative details the same. And while the surface meaning is about keeping one's neighbours' activities from marring the aesthetics and order of the neighbourhood, Rebecca reiterates that it is ultimately about maintaining housing values for resale, just as Perin predicted.

The economics of private governance: San Antonio, Texas

Gated-community debates frequently espouse overly simplistic notions of the private and public realms. Adding a third category – the 'club' realm – modifies this dichotomy (Webster 2001, 2002). From the point of view of public-goods theory, the consumption of a good or service is what makes it public or private. Cities or suburbs, in this scheme, are made up of small publics, each of which may be thought of as a collective consumption club (Webster 2002).

These 'consumption clubs' assign legal property rights over neighbourhood public goods by property market institutions (ownership, real-estate assessments) and are not different from other governance institutions where included 'members' (taxpayers) share goods and 'non-members' (non-taxpayers) are excluded. A comparison can be made between municipal governments supplying collective services on the basis of taxes, and gated communities with their own management companies, local service budgets and contractual regulations providing goods and services based on fees. Rather than focusing on the distinction between public and private, Webster argues it is more fruitful to ask whether gated communities deliver more equitable urban services to 'club' residents (Webster 2001).

The current growth in propriety communities is a response to municipal government's failure to provide adequate neighbourhood services because of 'free-riding' – that is, services being used by those who do not pay for them – and because of the inability of local governments to supply services in rapidly growing areas. Instead, by 'bundling up' a variety of public goods within a residential scheme and recovering costs through sale prices and fees, the house-building industry and market becomes a neighbourhood-building industry and market (Webster 2001: 163). For the people living inside gated communities it is an efficient solution, because of the legal requirement to pay fees and the homogeneity of community needs and desires, and would-be residents can choose their package of communal goods according to their personal preferences (Webster 2001: 164). The Mitchells, who retired to Sun Meadow, illustrate this point. They lived in an incorporated town within San Antonio where they liked the schools, had their own police department and purchased services from the city. They found the services supplied by San Antonio adequate but expensive, and therefore decided to move to Sun Meadow where services were initially provided by the homeowners' association.

Economic recession and the decline of real-estate prices, however, also altered the level of services and tax structure, particularly in Texas. Pamela and Wayne, who live in the Links, offer their perspective on how taxation, private governance and land-planning influence fees and services in their development and other gated communities.

Pamela and Wayne – municipal planning and gated communities

Pamela and Wayne live in a Florentine mansion just inside the guardhouse at the Links. The entry hall is set with inlaid marble and stone pillars, furnished with French Empire antiques. Pamela is tall and slim with long blond hair, wearing white satin lounging pants and a silk cowboy shirt. Wayne towers above her, casually dressed for work on a construction site. We met at a dinner party where I asked if I could interview them.

Their section of the development was planned with well-to-do retirees in mind and mainly doctors, lawyers and other professional retired people live in their immediate neighbourhood. Most of their neighbours moved from an even bigger home, one because he plays golf, and another because she wants to be close to a hospital. The majority of their neighbours are white, but also include a retired black general, a black insurance agent, and Mexican nationals, who 'keep a shopping home here'.

Wayne is familiar with the municipal consequences of developing large-scale gated communities outside the city in terms of metropolitan planning and governance, and the ability to provide city services to communities:

> A lot of what's happening in these gated communities is that local government continues to annex.[4] As they annex, the level of service

> goes down, not only for the annexed areas, but also for the inner parts of the city. [The idea is that] we'll get more money and everybody will benefit. [But in fact] they get more taxes and dilute the level or quality of services.
>
> So developers say, 'well, I have got to have my own amenity package, I have to have my own gated community'. All that is, is a reflection of the inability of the government to deal properly with the problems at hand. They create another level of service: homeowners' associations in these gated communities. So the type of development we get is because the city is ambitious in terms of annexation.
>
> I see problems with gated communities in the future. For instance, when you put in an asphalt street and you limit the driving on that asphalt street to just the people who live in that subdivision, you're not getting a lot of traffic. Asphalt streets deteriorate without use. But the reality is that when gated communities are annexed, the city does not take responsibility for repair of those streets. Homeowners' associations have to provide that. And in areas where the access is restricted, the asphalt is cracking and deteriorating because no one's driving on it.

Wayne believes that the community will be annexed within two years. It means that residents will have to pay city and county taxes as well as their homeowners' association maintenance fees. Tax assessments will also increase. But the city will not take over responsibility if they are private streets, or anywhere it is not public. He explains that the city does offer some services such as police and fire protection, but because the community is gated, there are still problems of access. According to him, homeowners' associations in gated communities are setting back San Antonio's ability to have proper levels of municipal services with appropriate taxation.

For Wayne one of the biggest drawbacks to gated communities is what is going to happen with taxation, annexation and the relationship with the city and with municipal governments. The critical issue is the provision of city services and equity of taxation. He feels there ought to be an equitable breakdown as to what you pay for and what you receive. 'And when these private communities get annexed, they're going to be paying an exorbitant amount of tax for what benefit they get, and the inequity gets even further out of line.'

Gilbert and Marie – taxation and provision of services

Taxation and provision of city services are also important to Gilbert. He moved to Sun Meadow because it is a private community that provides its own services at a reasonable cost.

Gilbert and Marie Walbaum live in a split-level, brick ranch with a three-car garage big enough for their golf carts. They retired in 1989, moving first to Florida and then to San Antonio. They are both still active – Marie is 69 and

Gilbert is 71 – and involved in the homeowners' association, the bridge club and their church.

They initially retired to a subdivision near a town of 5,000 people in Florida. He was active in the homeowners' association, where property owners already had control and decision-making was easy. In Sun Meadow the developers still own a majority of the lots, so working on the homeowners' association is much more frustrating. They left their first retirement community because it was never completed and not gated.

> I ask him about Sun Meadow:
>
> I wanted a clean, safe place to retire. The country has changed. It didn't affect me too much, but it affected some friends of mine. I felt bad, and I don't like some things about gated communities. Gated communities are not treated properly by the government. But it keeps them out of my business a little bit, so maybe it's worth it. You know streets are a large part of the city. If they don't have to build them and maintain them, then that's big money . . . What I'm saying is [what] bothers me is the make-up of the city and how that city is governed, the reactions that the city directors give.

Gilbert thinks that gated communities are not treated well in terms of taxation. Part of his motivation for moving to a gated community is that there is a private government, a homeowners' association, providing services within the development. As he put it: 'Any time I can keep the government out of my business I do.' He thought being in a gated community would keep the government out, but now Sun Meadow is to be annexed by the city. When he purchased his home, they were two and half miles outside the city limit. But now 'we're going to be a gated neighborhood in the city'.

For Gilbert the idea of having a private government that residents control is the most important part of the gated community. The Mitchells, on the other hand, consider the provision of adequate services the most important issue, and Wayne and Pamela argue that it is San Antonio's lack of municipal planning, rampant annexation to increase the tax base, and resulting inequality in rates of taxation in relation to the provision of services.

These issues are not mentioned by Manor House residents because of differences in the structure and stability of local government. Manor House, like Sun Meadow, is located just outside the city border, but New York City is no longer expanding by annexation and has not since 1898. The level of taxation and provision of services is determined by the local township and county and did not change during the duration of this research. Based on these interviews and others similar in content, then, Webster's theory of private governance for economic efficiency and the provision of services is substantiated in the Texas gated community examples, because of the rapid growth of the city, but not in the New York region cases.

Creating their own residents' association: Mexico City

The case of the Mexico City gated community, however, suggests that providing adequate security and services, such as maintenance of the roads and cleaning up the landscape, may become so important that residents in a suburban development will create their own form of private governance even without the legal underpinnings of common-interest developments or the support of the municipality or state.

Maria Regina – inadequate security and maintenance

Maria Regina, the president of the residents' association at Vista Mar, is in her early thirties, dressed in a white silk blouse and skirt. She lives in a modern, one-storey stone and glass house with her husband and son, and a maid. She begins by saying that there are subtle class differences in the community:

> Some are decent people with more or less economic power. We have a lot of new rich with inferior cultural background, and this may cause some problems . . . They have jobs in the *central de abastos* (market and distribution centre) or they are truckers. They have money but not education and this can cause problems with those that may have not had a lot of money, but are educated.'[5]

She later attributes these class and cultural differences to differences in participation in the community. But she gets along with everyone, and has a group of friends that she visits. It is a good relationship, because she is free to stop by for coffee or to borrow an egg, and you can ask a neighbour to watch your house while you are away.

Maria Regina explains that in Mexico City gated communities are not formed as common-interest developments but as private subdivisions with lots purchased individually without any guarantees or a required fee structure. At Vista Mar a team of seven women are in charge of the administration of the residents' association and take care of maintenance as well as relations between the neighbours and the authorities.

Since the beginning Vista Mar has been closed by four gates. The gates are illegal as it is not constitutional to restrict access to public roads, but people are gating their communities throughout their area because the state cannot provide adequate security. Gating has a cost for the residents, since the state will not provide maintenance for the streets or police surveillance in a private area. For this reason the residents' association imposed a maintenance fee even though they do not have the legal power to force residents to pay. Part of their work is to convince people that if they want to live in a gated community they have to pay a fee to have the necessary services. Only about 45 per cent of the residents pay the fees – she attributes this to cultural differences – so of the four entrances, one is reserved only for the people who pay. The fees also cover the three parties a year that are attended by everyone, whether they have paid or not.

Maria Regina's team oversees the maintenance of the million square metres of the entire community and 120,000 square metres of garden areas. They also clean the 200 vacant lots to reduce pests and limit security problems. They provide lighting and water services and charge to bring water to the houses since the community is not yet connected to the city water network. They coordinate the gardeners, office workers and policemen, and hire the security employees.

Residents have been ignoring the restrictions that were set up originally. Because Vista Mar was built in a semi-ecological zone, the developers were only allowed to construct houses with significant green areas, and houses could not exceed a specific size. But unfortunately builders were violating these internal regulations because they were not codified in state or municipal law. The women's group was concerned that if they did not take over the administration and begin to enforce the regulations, they would lose the gardens that were so important to the women and children.

So even when there was no supporting legal structure such as CID, or other form of governance other than state regulations for the subdivision development, residents created a quasi-governmental system to handle the administration of the community, to collect fees for services and to make decisions that benefit everyone. Even if only half of the community pays the fee, it is adequate to maintain the grounds and hire security. No one even mentioned the existence of rules and regulations for house design, much less as a moral code, and house design within the community varies considerably. Instead class and cultural designations – new rich, decent people – pervade the interviewees, categories that are dependent on intrinsic characteristics of people's background or aesthetic taste that work to sort residents into their 'proper' places.

Conclusion

When examined from the New York residents' point of view, private governance minimises conflict by setting up an a priori set of rules and regulations that determines what residents can and cannot do with their property. It also provides a strict standard for the appearance of their homes and gardens, reducing the possibility of disagreement over whether a neighbour did or did not mow their grass or put their garbage out too early in the day. It reinforces suburban moral minimalism by predetermining who is willing to live with the level of rules and regulations imposed by the homeowners' association.

In Texas, on the other hand, private governance from the residents' point of view is about the efficient provision of services, and in Mexico City this form of governance becomes a vehicle for collecting fees that can be used to provide security and maintenance not provided by the state. From these three ethnographic examples it seems that no one theory predicts the emergence of this governance form, but that local context and politics, the history of the community and urban growth influence residents' decisions and satisfaction with their choice.

Notes

1 From a series of interviews by Jörg Plöger completed in 2001.
2 These prices are for 1995–6. In 2001 it was reported by Jörg Plöger that a house sold for as much as 1.8 million.
3 Much of this information was gathered by Mariana Diaz-Wionczek.
4 Annexation is a process by which a town, county or other governmental unit becomes part of the city, in this case the metropolitan region of San Antonio.
5 The translation of this interview was completed by Lymari Benitez, a bilingual graduate student in environmental psychology born in Puerto Rico.

Bibliography

Agnotti, Thomas (1997) 'A metropolis of enclaves: Image and reality in North America', *Urbana* 22: 13–24.

Baumgartner, M.P. (1988) *The Moral Order of the Suburb*, New York: Oxford University Press.

Denton, N. (1994) 'Are African-Americans still hypersegregated?', in R. Bullard, J.E. Grigsby III and C. Lee (eds) *Residential Apartheid*, Los Angeles: Center for Afro-American Studies: 49–81.

Graham, L. (1999) 'Straddling the fence', *Dallas Morning News*, 7 August: 1.

Judd, D. (1995) 'The rise of new walled cities', in H. Ligget and D.C. Perry (eds) *Spatial Practices*, Thousand Oaks, CA: Sage Publications: 144–65.

Low, S.M. (2003) *Behind the Gates: Life, Security, and the Pursuit of Happiness in Fortress America*, New York and London: Routledge.

MacCormack, J. (2000) 'District 6 faces its own great divide', *San Antonio Express-News*, 22 May: 1.

McKenzie, E. (1994) *Privatopia*, New Haven, CT: Yale University Press.

Merry, S.E. (1993) 'Mending walls and building fences: Constructing the private neighborhood', *Journal of Legal Pluralism & Unofficial Law*, 33: 71–90.

Muller, P.O. (1996) 'The evolution of the American suburbs', in N.L. Shumsky (ed.) *Urbanization and the Growth of Cities*, New York: Garland Publishing: 395–404.

Nelson, R. (forthcoming) *Privatizing the Neighborhood: A Social Revolution in America.*

Perin, C. (1977) *Everything in Its Place: Social Order and Land Use in America*, Princeton, NJ: Princeton University Press.

Plöger, J. (2002) 'Gated communities in North Hills: Analyse der Ursachen ihrer Verbreitung in einer Vorortgemeinde der Metropolregion New York', unpublished master's thesis, Institute of Geography, University of Hamburg.

Ross, M., Massaron, L., Smith, J. and Pritt, R.D. (1996) 'The zoning process', *Urban Lawyer*, 28 (4): 801–7.

Webster, C. (2001) 'Gated cities of tomorrow', *Town Planning Review*, 72 (2): 149–70.

—— (2002) 'Property rights and the public realm: gates, green belts and Gemeinschaft', *Environment and Planning B: Planning and Design*, 29 (3): 397–412.

5

PRIVATE GATED NEIGHBOURHOODS

A progressive trend in US urban development

Klaus Frantz

Abstract

In the last two and a half decades the emergence of gated residential areas has become a mass trend in US urban development. It is estimated that more than 8 million people live in these communities today. Throughout the country they have changed the urban landscape as well as suburban society and its lifestyle. Gated communities are one element in US cities that reflect increasing polarisation and fragmentation, and diminished solidarity, within urban society and the progressive trend towards privatisation of municipal services at a time when both the quality and quantity of these services have declined in most US cities due to reduced public spending. This chapter focuses on basic aspects and issues, such as the physical and functional characteristics of gated residential developments, where they are to be found and who lives there and why. Research was carried out in the metropolitan areas of Phoenix, Miami and Los Angeles, the main emphasis in this chapter being placed, however, on Phoenix, the capital of Arizona, where all the gated communities were mapped and the total population recorded.

Introduction

In the past 25 years gated residential developments, known as *gated communities*, have altered the American urban landscape as well as American suburban society and its lifestyle. Along with the emergence of these privately administered residential areas came a change in the quality and scope of public administration and services for the remaining urban areas. Both the quality and quantity of these municipal services have declined in most US cities owing to reduced public spending, while the private sector specialising in these new communities has boomed. The residents and elected representatives of these communities could not manage without this private service sector industry.

In view of these fundamental changes in the urban landscape, it is therefore surprising that neither American nor any other urban researchers have dealt with this phenomenon for very long. Only in the last few years have studies on gated communities been conducted and these were predominantly written by urban planners, political scientists and anthropologists (see Davis 1990; Flusty 1994; McKenzie 1994; Caldeira 1996; Blakely and Snyder 1997; Marcuse 1997; Stark 1998; Low 2003). However, as very little systematic fieldwork has been done on the subject, urban research has a great deal to catch up on.

The intention of this chapter is to focus on basic aspects and issues, such as the physical and functional characteristics of these gated residential developments, where they are to be found, how widespread this new type of settlement is and who lives there and why. Research was carried out in the metropolitan areas of Phoenix, Miami and Los Angeles, the main emphasis in this chapter being placed, however, on Phoenix, the capital of Arizona. Here all the gated communities were mapped and the total population recorded in what may be one of the few city-level enumerations of the phenomenon.

Physical and functional characteristics of gated communities

A design of protection and fear

Gated communities are mostly private residential settlements whose residents cut themselves off from the outside world by using a whole series of defensive means, both of an architectural and of a landscaping nature. These measures include gates that are guarded day and night or gates that only open by remote control (figure 5.1). Sometimes these closed communities are protected by walls, fences or earth banks covered with bushes and shrubs, though this type of protection often only occurs on the outside of the parts of the community that could otherwise be seen from the road. Protection is also given by inaccessible land or vast expanses of land such as a nature reserve, and sometimes there are moats and guarded bridges. In addition, high-tech surveillance systems (monitors, night vision aids or radar screens) are used, which resemble the equipment of military security plants.

In the immediate vicinity of gated communities one often comes across a kind of 'landscaping of fear': devices of security and deterrence, which make it difficult for people to approach these settlements from the outside. For example, there are very few sidewalks close to gated communities. One cannot even park one's car, or stop for a short time, nor are there any signposts showing the way to many of these communities.

Inside the settlements, protection is guaranteed either by a so-called neighbourhood watch or by professional security personnel that patrol around the clock.

Figure 5.1 The artificial lake at Westlake gated community in the north-west of Los Angeles County, 2002. Photo: Klaus Frantz

Types of housing

Communities consist of various housing mixes including single family, townhouses, owner occupied, rented apartments and even mobile homes. Gated communities may include all of these types of home or just a single type. As far as size is concerned, communities may consist of just a handful of houses, provided the residents have the financial means to maintain the infrastructure, or there may be as many as 15,000 homes, as is the case in Laguna Hills, one of the eight US 'Leisure Worlds'.

Leisure World Laguna Hills in Orange County, in the south of the metropolitan area of Los Angeles, is the largest gated community in North America, being a retirement community with over 22,000 residents. Founded in 1964 and gaining city status in 1999, this community, with its 14 guarded gates, has succeeded in remaining independent and no longer has to make fiscal contributions to neighbouring communities, for example to support the local school district. 'Why should we pay taxes for schools if we don't need them here', is an argument frequently heard in retirement communities.

Facilities and amenities

Many gated communities are not simply a place to reside, but places where residents can enjoy leisure activities in an undisturbed private atmosphere and where they can follow a lifestyle that they have deliberately chosen. The people

Table 5.1 Gated communities within cities and towns of Metro Phoenix, both completed and under construction, and their leisure-oriented amenities (in %): 1999

	Number of inhabitants (1998)	*Total number of GCs*	*Still under construction (in %)*	*Access to golf course*	*Artificial lake (in %)*	*Adjacent to protected area (in %)*
Apache Junction	22,863	17	–	5.9	–	–
Avondale	28,650	–	–	–	–	–
Buckeye	5,035	–	–	–	–	–
Carefree	2,615	14	–	14.3	–	14.3
Cave Creek	3,605	2	–	50.0	–	–
Chandler	160,165	25	12.0	56.0	44	–
El Mirage	5,855	2	50.0	50.0	–	–
Fountain Hills	17,280	18	50.0	61.1	–	27.8
Gilbert	91,290	12	8.3	8.3	75.0	–
Glendale	196,820	20	–	15.0	15.0	–
Goodyear	14,305	3	33.3	00.0	33.3	–
Guadalupe	5,390	–	–	–	–	–
Litchfield Park	3,870	1	–	100.0	–	–
Mesa	361,895	94	26.6	21.3	5.3	5.3
Paradise Valley	13,315	18	–	55.6	–	5.6
Peoria	89,930	11	18.2	36.4	–	9.1
Phoenix	1,220,710	149	6.7	27.5	1.3	22.1
Queen Creek	3,445	–	–	–	–	–
Scottsdale	195,495	214	30.8	35.5	7.5	16.4
Surprise	18,830	7	42.9	57.1	57.1	–
Tempe	159,220	34	–	–	2.9	–
Tolleson	4,675	–	–	–	–	–
Youngtown	2,780	–	–	–	–	–
Total	2,749,453	641	19.0	30.0	8.1	12.8

Source: author's survey

who market this type of community often advertise them as so-called *lifestyle communities*.

On the one hand the amenities advertised are those that nature itself provides: a favourable site, an attractive, unspoiled landscape right outside your front door. Your home may be close to the beach or adjacent to a protected area. By law it is illegal to cut off access to a beach or a nature reserve, but in reality gated communities occasionally hinder free access to such areas. On the other hand, amenities are constructed by the developers. Developers plan artificial lakes (figure 5.1) suitable for fishing, boating or water skiing, or simply to look good and raise property values. There are also parks complete with cycle paths or footpaths, tennis courts and golf courses. In Metro Phoenix, about 13 per cent of all gated communities are located adjacent to a protected area, 8 per cent have their own artificial lake; and almost one-third of them have access to a golf course (table 5.1). Swimming pools, fitness centres and country clubs are often also part of these private communities. There is no limit to the community focus designed into such schemes and developments can be found, for example, with polo courts, a private airstrip (figure 5.2) and bridle paths (figure 5.3). These facilities are sometimes at the disposal of several bordering private communities so that the costs for construction and maintenance can be shared, as is the case with the golf course at Airpark Estates, shared between three adjacent private neighbourhoods. Commercial or public facilities are rare in North American gated communities because that would necessitate giving access to non-residents. The insularity of private estates might be a distinctive feature of North American schemes. When the Hidden Hills gated community succeeded in becoming an incorporated city in 1961 the homeowners' association voted to ban outsiders from general access (figure 5.3 on p. 70). Hidden Hills, with its approximately 2,500 residents, is a so-called equestrian community that prides itself on owning an attractive, 42-mile network of bridle paths.

Gates within gates

It is not a rare phenomenon for several gated communities with varying socioeconomic status and distinguishable house styles to be located within one large gated master-planned community (MPC) scheme. The search for exclusivity, security and undisturbed peace does not apparently halt behind one set of closed gates. In the metropolitan area of Phoenix, as many as 60 gated communities lie within the walls and fences of other gated communities. The best example is Desert Mountain, a large, rigorously guarded and protected gated master-planned community in the north of Scottsdale, a city of about 195,000 inhabitants in Metro Phoenix. Desert Mountain, an exclusive MPC consisting of residential areas only, presently encompasses 29 gated communities with a total of about 900 custom homes and 2,400 residents (Frantz 2001: 17).

Where are gated communities to be found?

In the US one comes across gated communities in all metropolitan areas. There are large numbers in the urban areas of the Sunbelt states, but they can also be found in attractive rural areas that have become popular as second-home resorts. In cities, gated communities are mostly located in newly developed districts, typically characterised by a winding circular road network with many dead ends. Private communities – gated or ungated – usually have only one or two access roads as this facilitates security patrolling. In older districts, where a grid pattern predominates and roads are publicly owned and maintained, it would be more costly and difficult to close off residential areas to the public. Having said this, there still are examples of gated communities in such areas.

A good example of a city with a mixture of old and new street networks and a corresponding pattern of gated communities is Coral Gables in Metro Miami, an MPC from the 1920s with 42,000 inhabitants. In the northern,

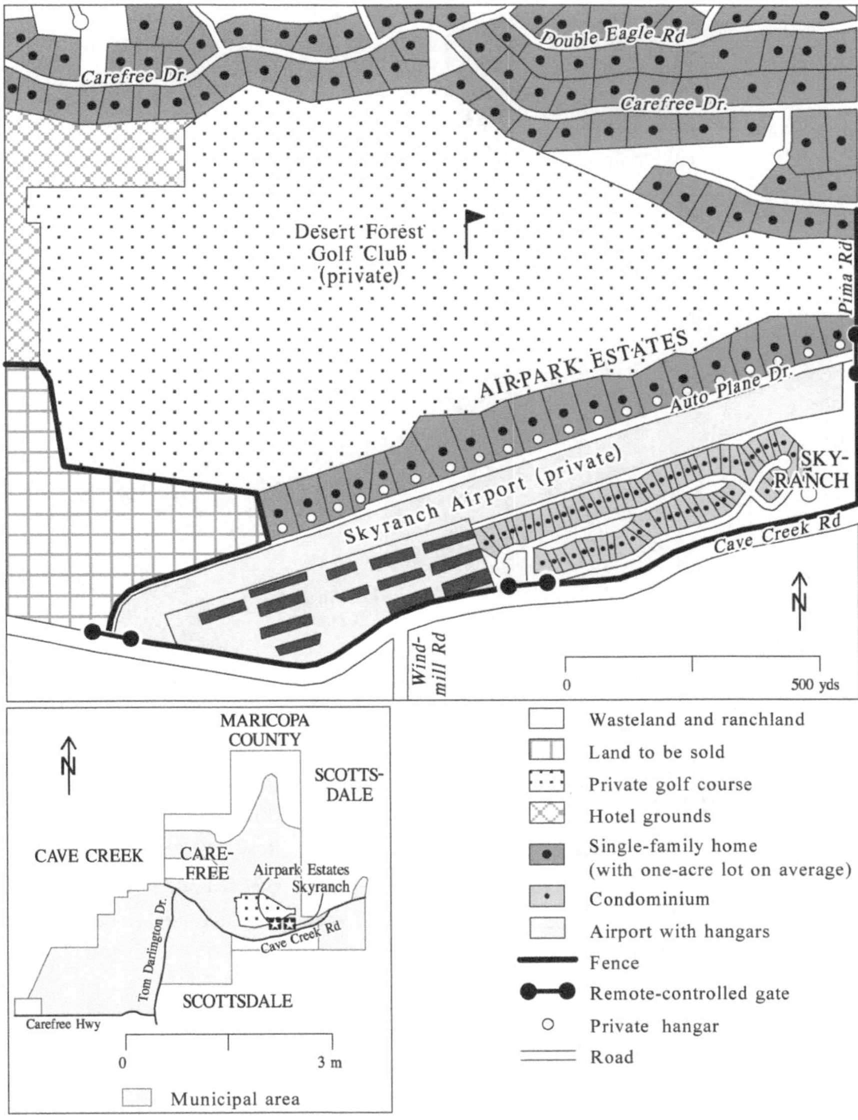

Figure 5.2 The gated communities Airpark Estates and Skyranch in Carefree – in Metro Phoenix, Arizona, 2000. Source: author's enquiries

older parts of town that have a grid pattern of roads, there are no gated communities. One does, however, come across street closures, which have been set up in the last few years, as in other US cities (see Jones and Lowrey 1995). The affluent people living in this part of town have barricaded themselves behind gates and built constructions on public roads, blocking off the through traffic. In the adjacent, somewhat run-down immigrant districts of the city of Miami

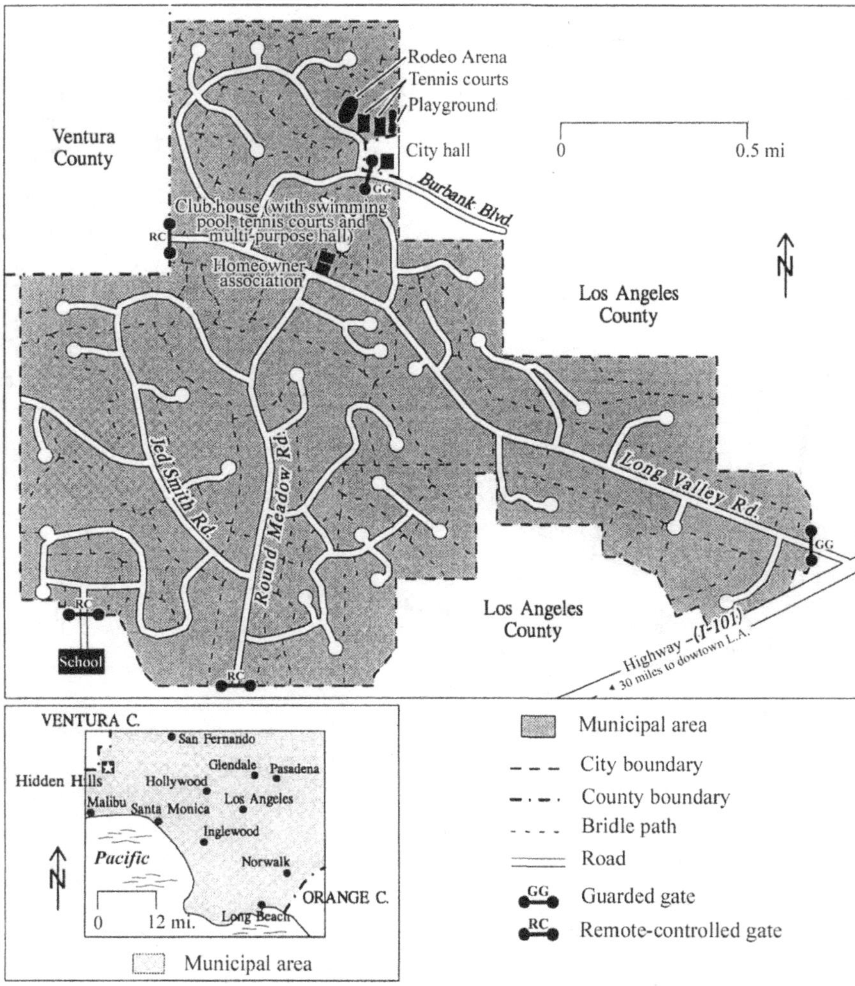

Figure 5.3 Hidden Hills in Los Angeles County, California: a gated, private city, 2000. Source: author's enquiries

to the north, and West Miami to the west, populated by Hispanics, the residents dislike street closures. In contrast to the antagonist in Robert Frost's poem they do not believe that 'good fences make good neighbours' ('Mending Wall', in Frost 1946). The reason for these street closures is that traffic is diverted from the neighbouring enclosures through their residential area.

In the southern, more recently built parts of the city where the road network is curved, there are as many as 13 gated communities with one or two access roads each. The ten oldest of them still have a road network, financed and maintained by public tax revenues. The guards at the gates of these developments cannot therefore prevent outsiders from entering. They do, however, ask to see an identity card. In the three most recent gated communities, Tahiti Beach, Journey's End and Deering Bay, where the roads are financed and maintained privately, outsiders may not enter without an invitation.

The number and size of gated communities and the number and characteristics of people living there

In the mid-1990s the two American urban planners Blakely and Snyder estimated that in the United States more than 8 million people were living in gated communities (1997: 180). However, as there are no official statistics available, the current numbers of people living in gated communities today remain unknown. To gain a more precise picture of this new type of settlement, first-hand statistics were created by the author by undertaking an extensive and detailed survey of a single US metropolitan area. For this purpose Phoenix, the fastest growing US city of more than a million people, was chosen.

In this sprawling metropolitan area of about 2.8 million people, the author systematically searched for gated communities. To do this, the entire area of Metro Phoenix was covered by car in order to find the locations of all gated communities, to analyse each one of them, and to build up a database. In addition, town planners, policemen and firefighters from each of the 23 independent cities and towns were interviewed. In case of emergency, firefighters have a special key to open the remote-controlled gates of the communities which are made up of condominiums only, thus they know the exact location of such neighbourhoods.

Employees at real-estate offices and construction companies, as well as a large number of residents and board members of homeowners' associations, were also consulted. The information obtained was used to ascertain the socio-economic status of gated-community residents and to suggest reasons why gated communities are in such demand.

In the US the price of a house or condominium is a good indicator of the socio-economic status of a neighbourhood. The price can be estimated by looking at a number of factors, such as the type, size, quality and location of a house. In some cases county property appraisers were consulted who calculate the (taxable) value of each property within a neighbourhood. The assessment rolls in city halls and county seats also list the assessed price of houses.

Phoenix – a case study

The results of the survey carried out in 1999 showed that there are altogether 641 gated communities distributed over 17 of the 23 municipalities within Metro Phoenix (table 5.1) – 214 of these are in Scottsdale, 149 in the city of Phoenix and 94 in Mesa. Of these communities, 19 per cent were being constructed at the time of the survey, which illustrates how fast this new type of settlement is spreading. Guarded gates were found in 19.3 per cent of communities and 11.9 per cent were retirement communities.

The majority of gated communities in Metro Phoenix are small residential areas. Almost 50 per cent of them consist of fewer than 100 housing units. Only 1 per cent of the communities have 2,000 or more housing units. These larger developments are mainly retirement communities with mobile homes.

Approximately 320,000 people are estimated to live in Phoenix's gated communities. This is 11.7 per cent of the total population of Metro Phoenix. However, the proportion of people living behind gates varies considerably within the city. In Surprise, a city of 19,000, almost 70 per cent of the population live in gated communities. Nearly half of all the people in Apache Junction and Carefree, and 42 per cent of those in Scottsdale, whose total population counts 195,000, live within gates. In all these cities the number of retirement communities is high. Almost 29 per cent of all the residents in the gated communities of Metro Phoenix live in retirement communities.

Metro Phoenix demonstrates that no longer only the wealthy live in gated communities. No more than about 13 per cent of all the closed residential areas are enclaves of the upper class. Almost 49 per cent are communities of the middle class and approximately 2 per cent are home to lower middle-class families (Frantz 2001: 16 and 18).

Reasons for living in a gated community

Desire for security and fear of crime

The most obvious reason for moving to a gated community is the fear of increasing crime and the citizens' desire to feel safer (Marcuse 1997). Undoubtedly, the crime rate in US cities is high in comparison with the situation in central Europe, but paradoxically, at the same time as gated communities have been springing up all over the United States in the past 25 years, the crime rate in the country has dropped. It may also be noted that most of the gated communities are located in those parts of town where the crime rate has always been very low. The residents' fear of a high crime rate may, therefore, be real enough, but relatively unfounded. This mirrors the situation in many European countries where the number of people who report having personally experienced a crime against them has been falling since the mid-1990s, while fear of crime in the same period has been rising. Gates, walls and fences cannot stop uninvited people from entering a community if they have set their minds on doing so. It might be supposed that such fortifications inhibit opportunistic crime – as well, of course, as displacing it. It may well be that security accoutrements contribute to a heightened sense of security perceived by the residents, and that residents are relatively insensitive to any actual effects on real crime.

Protection of the private sphere

Gated communities do not permit through traffic. This is how they deter opportunist crime since petty thieving is often a vehicle-based business. Gated communities present carefree environments for parents who elsewhere would have to worry about traffic, abductions and the character of their children's playmates.

Exclusivity

The exclusivity and status of these closed communities is often expressed in majestic gates and driveways, resembling the driveways and gates of Old World mansions that belong to the aristocracy. These status symbols that exist in a good many closed residential areas satisfy the search for exclusivity and provide prestige status to residents. The accoutrements of exclusivity are equally prominent in the marketing material of developers and real-estate agents and add value to their products and add resale value for home investors. Anecdotally there is evidence that houses sell more quickly and at a higher price when built in a community with fortification and design prestige. However, there are as yet few rigorous empirical studies to verify this claim.

Predictability and property values

Every homeowner in a gated community has to sign covenants, restrictions and conditions when purchasing deeds. Thus the residents are obliged to abide by a pre-understood set of behavioural rules, which are understood by all to be necessary for the maintenance of the quality of the community. Because of this binding legal framework and the methods used to enforce these rules, houses in (gated) common-interest developments should not, in principle, be so vulnerable to value fluctuation as conventional communities where resident profile, land use and local externalities are less predictable. In principle, this makes privately governed communities a less risky investment.

Retreat from failing public government

Citizens' dissatisfaction with the politics and the services of public municipal government is widely reported in popular and academic discussions about gating. This is a plausible contributory explanation for the rise in the number of people who are opting into private neighbourhoods. Homeowner associations, independent of municipal politics, provide a sort of shadow government, locally more responsive (though not without local politics) and better able to supply services of a quality and quantity demanded by community residents.

Identity, packaging and social homogeneity

Gated communities are each distinguished by a certain architectural style and by a package of facilities and amenities designed to appeal to a certain niche and to compare favourably with other communities in the vicinity. The packaging of neighbourhood life is emphasised in the names given to these settlements:

Hidden Paradise, Palm Gardens or Desert Diamond, for example, in Metro Phoenix. Developers are trying to create a sense of exclusivity but are also signalling the particular mix of qualities that distinguishes a scheme from competitor schemes. Important in this respect is the profile of neighbours that buyers can expect to find after they move in – most explicit in the case of age-specific common-interest schemes. All these features create a specific identity, making a scheme stand out from undifferentiated traditional American suburbia. Gated communities not only give many residents a feeling of security, but also make them feel at home in sprawling metropolitan areas where communities often look alike and neighbours come and go and often have nothing to do with each other. In searching for particular selling points and creating new housing market niches, developers are therefore from one point of view helping differentiate suburbia. They may also be helping to fragment it in ways that are less benign.

Conclusion

The emergence and rapid spread of gated communities reflect and strengthen the trend in America to privatise urban space and public services. The popularity of privately administered residential areas means increasing differentiation and associated fragmentation, segregation and individualisation of urban society. This is all incontrovertible. What is contested – and widely discussed in the literature on gated communities – are the effects such increasing spatial differentiation will have on social polarisation and a diminishing feeling of solidarity. Gated communities are such a fundamental break with the traditional pattern of boundary demarcation that they are rightly viewed with caution and suspicion by those taking the longer-term view. They clearly deliver great short-term efficiencies to many – why else would intelligent and consenting adults move into them in such large numbers? – but they might also take us a quantum leap forward in the fragmentation of urban space.

Gates and walls help establish a community's unique identity and define and reinforce the boundaries with 'other', especially less desirable, neighbourhoods. They also function as selling tools for the real-estate industry and as a symbol of prestige for the people living there. Gates and walls, however, are not just a statement about identity and status, but also about society's perception of crime, risk to person and property and search for security. There is a practical and material self-interest in choosing to live in a fortified enclave. More and more Americans are buying property here because they believe that they themselves have the power to control and maintain the physical and social quality of their neighbourhood. They feel that this is a way to guarantee that the community they have bought themselves into will not change drastically in the course of time – as it does in so many suburban residential areas of the US.

Bibliography

Blakely, E.J. and Snyder, M.G. (1997) *Fortress America: Gated Communities in the United States*, Washington, DC and Cambridge, MA: Brookings Institution Press and Lincoln Institute of Land Policy.

Caldeira, T. (1996) 'Fortified enclaves: The new urban segregation', *Public Culture*, 8: 303–28.

Davis, M. (1990) *City of Quartz: Excavating the Future in Los Angeles*, New York: Verso (The Haymarket Series).

Flusty, S. (1994) *Building Paranoia: The Proliferation of Interdictory Space and the Erosion of Spatial Justice*, West Hollywood: Los Angeles Forum for Architecture and Urban Design (Forum Publication 11).

Frantz, K. (2001) 'Gated communities in Metro Phoenix (Arizona). Neuer Trend in der US-amerikanischen Stadtlandschaft', *Geographische Rundschau*, 53: 12–18.

Frost, R. (1946) *Collected Poems by Robert Frost*, New York: Random House (The Modern Library).

Jones, M. and Lowrey, K. (1995) 'Street barriers in American cities', *Urban Geography*, 16: 112–22.

Low, S.M. (2003) *Behind the Gates: Life, Security, and the Pursuit of Happiness in Fortress America*, New York and London: Routledge.

McKenzie, E. (1994) *Privatopia: Homeowner Associations and the Rise of Residential Private Government*, New Haven, CT: Yale University Press.

Marcuse, P. (1997) 'Walls of fear and walls of support', in N. Ellin (ed.) *Architecture of Fear*, Princeton, NJ: Princeton Press: 101–14.

Stark, A. (1998) 'America, the gated?', *Wilson Quarterly*, 22: 50–8.

6

GATED COMMUNITIES AS PREDATORS OF PUBLIC RESOURCES

The outcomes of fading boundaries between private management and public authorities in southern California[1]

Renaud Le Goix

Abstract

Based on an empirical study of the Los Angeles region, this chapter focuses on the consequences of gated-community development, leading to fading boundaries between private and public management of residential areas. As a standardised form of urban product, these neighbourhoods represent a form of urbanism where public spaces are being privatised. In the most recently urbanised areas, they represent an increasing part of the new-homes market and they have thus become a symbol of contemporary metropolitan fragmentation and social segregation. They not only enclose space but also actively select residents through restrictive covenants as well as through lifestyle marketing and price. Because they are managed as private corporations, there is perhaps an inevitable tendency to seek political and fiscal independence through a process of municipal incorporation. This has led to a project of partition – strengthened by and strengthening existing partitioning movements – and to the prospect of increased social segregation. The outcome is the fading of boundaries between public and private management when a gated community engages in municipal incorporation. The sprawl of gated communities is not to be understood as secession from public authority, but as a public–private partnership. It is a local game where the gated community provides benefits to the public authority, in return for which the property owners' association is granted autonomous local governance. The spillover effect of this method of ordering new urban space is to increase segregation. This is particularly so when gated communities are incorporated since the municipal institution is instrumental in securing public funds and property for the privilege of a gated enclave.

Introduction

In terms of private urban governance the US has a wide range of experience, which allows investigation of the issues raised by their long-term implementation. As a specific form of private governance, walled and gated residential neighbourhoods have become a common feature within the fastest-growing US metropolitan areas. Numerous gated communities have been developed since the 1960s in southern California, which serves in this chapter as a field of study for discussing the consequences of fading boundaries between public and private management due to the rapid development of these enclaves. Because security systems and around-the-clock gates prevent public access, gated communities represent a form of urbanism where public space is effectively privatised – protected for the exclusive consumption of a spatially and legally defined group. They differ from condominiums and secured apartment complexes because they enclose more natural real estate (as opposed to land created by investment in high-rise development). In some cases this land and its public infrastructure, such as streets, parks, sidewalks and beaches, was formerly open to all. In some cases some of the infrastructure is still publicly owned. Gated neighbourhoods have greatly increased in number since the 1970s and have become a powerful symbol of the fragmentation and increasing social segregation of contemporary cities (Blakely and Snyder 1997). Social segregation goes hand in hand with developments like this because they are managed as private corporations, tend to seek political autonomy and practice implicit selection of residents.

In the social sciences literature about gated communities three types of argument are now part of the general theoretical discourse. First, gated communities are described as part of a trend towards the commoditisation of urban public space (Dear and Flusty 1998; Sorkin 1992). This is often linked with the spread of ideologies of fear developed by economic and political actors including municipalities, the home-building industry, the security industry and the media (Davis 1990, 1998; Flusty 1994; Marcuse 1997). A second type of argument presents gated communities as a symptom of urban pathologies, among which social exclusion is considered to be pre-eminent. In this discourse, the decline of public spaces in cities is seen as being detrimental to the poorest social classes, and voluntary gating is associated with increased social segregation (Blakely and Snyder 1997; Caldeira 2000; Webster, Glasze and Frantz 2002). The shift from a city with public spaces to an urbanisation formed of private enclaves, it is argued, is associated with the 'secession' of the elite and constitutes a regressive redistribution of welfare (Donzelot and Mongin 1999; Jaillet 1999; Reich 1991). These are largely equity arguments that stand opposed to the efficiency arguments for gating based on the assumption that the public provision of services leaves potential welfare gains unrealised (Foldvary 1994). The debate about gated enclaves has been lively, despite a lack of empirical studies. Due to the difficulty of gathering a representative sample of gated communities at the local scale, empirical studies in the US have so far focused on the nationwide phenomenon (Blakely and Snyder 1997), or on a narrow municipal scale and a few case studies (Bjarnason 2000; Lacour-Little and Malpezzi 2001).

Based on an empirical study in the Los Angeles region,[2] this chapter focuses on the diseconomies created by gated communities for neighbouring communities. It aims to identify the ways in which local governments favour the development of this form of land use to pay for the cost of urban sprawl. This issue is addressed by focusing on gated communities that have been part of an incorporation process – seeking municipality status and pushing for political autonomy. In this process, the boundaries between private interest and the public realm become considerably blurred. I argue that the incorporations of gated communities create spillover effects because public funds and property are captured for the privilege of a gated enclave.

Gated-community development as a public–private partnership

The gated-community lifestyle is clearly inspired by the historical golden ghettos found in industrial-era cities – in Paris, the Montretout gated neighbourhood in Saint-Cloud developed in 1832 and Llewellyn Park, New Jersey developed in 1854 near New York. Gated enclaves today, however, are mainly suburban neighbourhoods distinguished not so much by their grand idealistic designs but by their emphasis on 'community lifestyle' and security features. The promotion of these standardised commodities by the real-estate industry typically focuses on exclusiveness, protection of families in a secluded environment and leisure facilities and amenities such as golf courses, private beaches, private parks and horse-riding trails.

In Los Angeles the first gated neighbourhoods were developed in 1935 in Rolling Hills and in 1938 in Bradbury, and some well-known gated communities were built early after the Second World War. These include Hidden Hills (1950) and the original Leisure World at Seal Beach, housing veterans and the retired (1946). Before 1960, about 1,700 housing units were gated in the Los Angeles area. This increased to 19,900 in 1970 through the development of major enclaves like Leisure World (1965) and Canyon Lake (1968). After 1970 new developments were usually smaller and the growth rate decreased: 31,000 gated units existed in 1980, 53,000 in 1990, and 80,000 in 2000. In the year 2000, this market represented approximately 12 per cent of the new-homes market in southern California. Locally it is higher, however: 21 per cent in Orange County, 31 per cent in San Fernando Valley and 50 per cent in the desert resort area of Palm Springs.[3] As real-estate commodities, they are tailored to a specific prospective buyer profile. The gated communities surveyed in the Los Angeles study are located within every kind of middle-class and upper-class neighbourhood and are available for every market segment. Half of them are located within the rich, upper-end and mostly white neighbourhoods, while one-third are located within the middle-class, average income and white suburban neighbourhoods. As evidence of the social diffusion of the phenomenon, 20 per cent of the communities surveyed are located within average- and lower-income Asian or Hispanic neighbourhoods, especially in the northern part of Orange County and in the north of San Fernando Valley (Le Goix 2002, 2003a).

The diffusion of homogeneous residential suburban communities is related to suburban growth, to the anti-fiscal posture and to the municipal fragmentation dynamics that have affected the Los Angeles area since the 1950s – issues that have already been very well documented. In Los Angeles the anti-fiscal posture has been associated with the incorporation of numerous cities – the first of which was Lakewood (1954). These municipal incorporations were designed to avoid paying costly county property taxes, which after incorporation were replaced by lower city assessments and better local control over local development and other municipal affairs (Miller 1981). A second important development may be linked to the 1978 'taxpayers' revolt' – a homeowner-driven property tax rollback known as Proposition 13 (Purcell 1997). Passed in 1978, the Jarvis-Grann Initiative introduced a 1 per cent limit of the assessed value for property taxes; annual increase is allowed up to 2 per cent a year. This tax limitation increased the need for public governments to attract new residential developments, especially those that would bring wealthy taxpayers into their jurisdiction. This set the scene for gated communities to become the perfect 'cash cow' (McKenzie 1994). A third influence on the spatial diffusion of gated enclaves is the rapid growth of the Los Angeles area (figure 6.1), sustained by massive population flows driven to the Sunbelt cities during the 1980s (Frey 1993).

In this context, the peculiarity of gated communities compared to regular common-interest developments (CIDs) and master-planned communities lies in the gating. On the one hand, access control and security features represent a substantial cost for the homeowner, not only the capital cost of infrastructure, but also ongoing maintenance costs that otherwise would be borne by the wider taxpaying public. As compensation, the homeowner is granted private and exclusive access to sites and to formerly public spaces. The resultant exclusivity enhances locational rent and positively affects the property values (Lacour-Little and Malpezzi 2001; Le Goix 2002). On the other hand, it provides the public authorities with wealthy taxpayers at barely any cost (McKenzie 1994).

As a consequence, the sprawl of gated communities tends to blur the limit between the public realm of municipal governance and the private realm of homeowners' association management. Indeed, the status of a CID is instrumental in transferring the cost of urban sprawl from the public authority to the private developer and ultimately to the final homeowner (McKenzie 1994). The enclosure movement has presented many interesting dynamics with regard to the financial situation of public governments since the 1978 tax limitation. For example, to the extent that gating favours property values, it helps increase the property tax basis. The erection of gates also transfers the cost of maintaining urban infrastructure to the association and the homeowner.

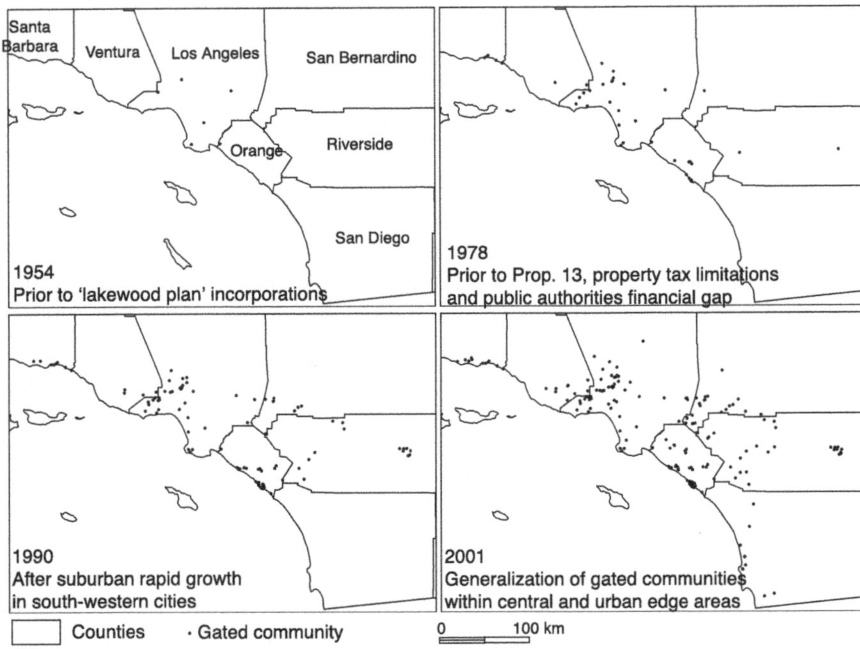

Figure 6.1 The diffusion of gated communities in the Los Angeles area. Sources: US Bureau of Census Boundary Files, database Gated communities UMR Géographie-cités/Le Goix 2002

The relationships between gated enclaves and the public authorities can be summarised thus: because of the fiscal gains they produce at almost no cost except general infrastructure (freeways and major infrastructure), gated communities are particularly desirable for local governments, especially in the unincorporated areas where budgets are tied to a low-resource paradigm after Proposition 13. The city of Calabasas offers a dramatic example of this when it incorporated in 1991. Incorporation is the legal process by which unincorporated land (under county jurisdiction) becomes a city, following approval by the state (in California, the LAFCOs, Local Agency Formation Commissions, are in charge of supervising the process) and by two-thirds of the voters. A new municipality can either be granted a charter by the state, as large cities are, or be incorporated under the general law, which is the common case. In Calabasas, at least 30 per cent of single housing units are located within gated enclaves. Baldwin, the developer of the first enclave in the area (Calabasas Park), proposed in 1987 a 2,000-unit extension on unincorporated land, while locally involved residents were pushing for municipal incorporation, willing to promote a slow-growth policy. This controversial drive towards local autonomy was aimed at gaining control of local land development and challenging the County Board of Supervisors to support new residential subdivisions. The incorporation process went back and forth for 11 years (as the LAFCO rejected it several times), while the county had already approved 4,500 new units in the area (Le Goix 2003b).

When developing private neighbourhoods, the homeowner pays for the provision of public services. The sprawl of gated communities is not, however, to be understood as 'secession' from the public authority, but as a public–private partnership, a local game where the gated community has utility for the public authority, while the property owners' association (POA) is granted autonomy in local governance, and especially in financing the maintenance of urban infrastructure. But this user-pays paradigm creates a high cost for the homeowner, charged with the property taxes, the district assessment and homeowners' fees. This higher 'entry fee' contributes to the protection of property values but also to socio-spatial selection and segregation.

Gated communities as municipalities and public actors

It has already been documented that common-interest developments (CIDs) are both public actors and private governments. The developer and the subsequent homeowners' association substitute for the public authority and privately provide a public service (Kennedy 1995; McKenzie 1994). But some gated communities have also become real public actors, by means of incorporation as autonomous cities, or by being a key actor in an incorporation process. This issue is important in understanding the nature of the territories built by gated enclaves. Large gated communities incorporated as cities in their own right include Bradbury and Rolling Hills (1957), Hidden Hills (1961), Canyon Lake (1991) and Leisure World (1999). Enclaves incorporated as part of a new city where a substantial part of single-family housing development is gated include Dana Point (1989), Calabasas (1991) and Dove Canyon (incorporated with Rancho Santa Margarita in 2000 – see table 6.1 overleaf).

Local affairs are shared between a private homeowners' association, in charge of road maintenance, security and compliance with land-use regulations and restrictive covenants, and a *minimal city*. Being a minimal city entails minimising the costs of operation by contracting with the county and other public agencies to supply public services (police, water, sewers and fire department) (Miller 1981). Indeed, the municipality in these incorporated areas acts as an extension of the property owners' association.

Generally speaking, such incorporations are conducted for two reasons, consistent with the global trend towards municipal autonomy in Los Angeles. On the one hand, they aim to prevent potential annexation by another and less affluent community looking for an extended tax base (Rolling Hills and Hidden Hills, for instance, or Rancho Mirage and Indian Wells in the Palm Springs area). On the other, they aim to promote the protection of a lifestyle and local values, and local control of affairs and planning (Leisure World, Canyon Lake, Dana Point, Calabasas).

First, having a municipality tailored to fit the needs of a POA is a common

Table 6.1 Gated communities involved in municipal incorporations

Municipalities	*Date*		*GC*	*Number of housing units*		*Units in GC*
	First GC	*Incorp-oration*		*in GC*	*in munici-pality (2000)*	*%*
Newport Beach	1968	1906	7	1,789	37,288	4.8
Rolling Hills	1936	1957	1	636	682	93.3
Bradbury	1950s	1957	1	476	1,261	100.0
Hidden Hills	1950	1961	1	592	592	100.0
Indian Wells	1957	1967	3	2,135	3,842	55.6
Rancho Mirage	1952	1973	2	4,122	11,816	34.9
La Quinta	1980s	1982	2	2,064	11,812	17.5
Dana Point	1975	1989	15	2,817	15,682	18.0
Laguna Niguel	1982	1989	2	1,664	23,885	7.0
Canyon Lake	1968	1991	1	4,047	4,047	100.0
Calabasas	1978	1991	6	2,228	7,426	30.0
Malibu	1975	1991	4	769	6,126	12.6
Leisure World/ Laguna Woods	1964	1999	1	12,736	11,699	100.0
Rancho Santa Margarita (Dove Canyon)	1986	2000	2	1,227	16,515	7.4

Notes
Number of housing units in gated communities are according to the database, and include units built and included in the assessment maps before 2000. Information based upon POAs data and/or assessment maps. Several contiguous country clubs or POAs, sharing the walls and entry gates, are considered as one unique gated community.

pattern of most gated-community incorporation, exemplified by Canyon Lake's incorporation in 1991 (9,900 permanent residents in 2000). The City of Canyon Lake operates according to the minimal-city paradigm. It is designed to provide residents with the basic services of police (contracted to the City of Perris police department), firefighting (contracted to the county fire department), sanitary infrastructure and zoning. Safety and security represent up to 61 per cent of the overall city budget (2000). All other services, including leisure, planning, development permits and regulations, are in the charge of the POA. In spite of its lack of power, the city acts as the public front office of the POA, especially when setting up zoning requirements. These seek strictly to avoid future development around the walls of the community. For instance, the municipality sought to acquire federal land on the outskirts of the city in 2000 in order to maintain a natural boundary between Canyon Lake ('a bit of paradise' – the official motto) and the sprawling nearby community (Tuscany Hills in Lake Elsinore).[4]

Second, should a gated community be concerned with the preservation of a certain lifestyle and identity, then incorporation is considered a means of gaining autonomy while avoiding annexation by less affluent municipalities seeking to enlarge their fiscal basis. Rolling Hills became the first gated community to incorporate as a city in its own right in 1957, and was soon joined in the move by Hidden Hills in 1961. Both are semi-rural up-market private estates, emphasising a ranching lifestyle, with horse stables and trails. Rolling Hills POA had already existed for two decades when the nearby middle-class and industrial Torrance City sought to annex the wealthy communities of the Palos Verde peninsula, south of Los Angeles (see figure 6.2 overleaf). Rolling Hills quickly reacted, filing for and voting for incorporation in less than 18 months (Miller 1981). The sprawling west side of the city of Los Angeles threatened Hidden Hills, which has strictly enforced a slow-growth policy since incorporation. In Rancho Mirage, also dubbed Country Clubs City (Palm Springs resort area) the wealthy retirement gated communities (35 per cent of the total housing stock and almost 65 per cent of detached units) led the process of incorporation in 1973 to avoid annexation by Cathedral City or Palm Desert.[5]

In some peculiar cases, external forces have convinced the POA to file an incorporation request. The incorporation of Leisure World as the City of Laguna Woods was an outcome of necessity. In the largest retirement gated community of the West Coast (19,500 inhabitants, whose average age is 77 years), incorporation became a necessity because of the lack of involvement in the operation of the Golden Rain Foundation (the master association of the community). As Robert Ring, the former president of the association's board of directors and now city council member, put it, residents 'don't buy green bananas . . . and don't bother as well, as they are happy'.[6] Several projects had been considered since 1964, but never passed. After Orange County filed for bankruptcy in 1995, the situation changed radically; the county promoted incorporation of urbanised areas and proposed that new minimal cities should use the local tax base to supply the residents with improved public services, and that they should sign contractual agreements with the county, which operates basic services (fire and sheriff's departments). Leisure World had to incorporate and there were several options. One was a joint incorporation with nearby communities housing younger populations in Laguna Hills or Mission Viejo. This option was rejected because of the obvious divergence of interest between a young population interested in schools and kindergartens and a retirement community. Another option was to be annexed by the large municipality of Irvine, but this gave rise to another conflict that helped to make the decision: Irvine supported a project for an international airport, the approach path to which would have flown over Leisure World (see figure 6.3 on p. 85). So, according to Robert Ring, the incorporation aimed to build a public entity that would be the voice and advocate of the Leisure World gated community against the international airport project in El Toro.

Finally, the main argument driving incorporation might be the need to retain the property tax dollars within the limits of a municipality. The incorporation of the city of Rancho Santa Margarita (2000) was complex; one large gated community was the main influence in this process (Dove Canyon, 1,230 properties), whereas its neighbour (Coto de Caza, 4,152 units) rejected the plan and remained unincorporated (it now wishes to build a municipality of its own). As in some other places, the incorporation aimed at gaining local control

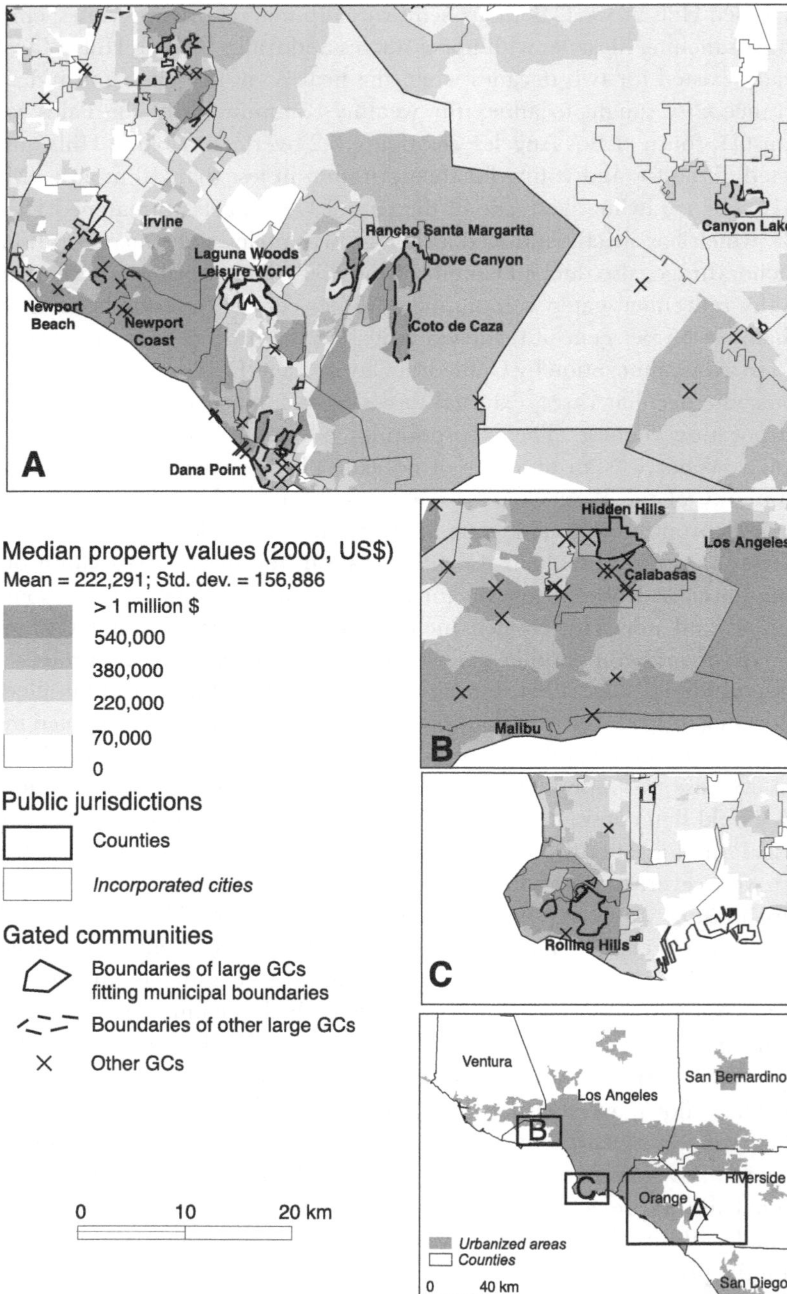

Figure 6.2 Gated communities, municipalities and property-value patterns in the Los Angeles region (focus on Orange County, Palos Verdes Peninsula and Western Los Angeles County). Sources: database Gated Communities UMR Géographie-cités 8504 2001, US Bureau of Census

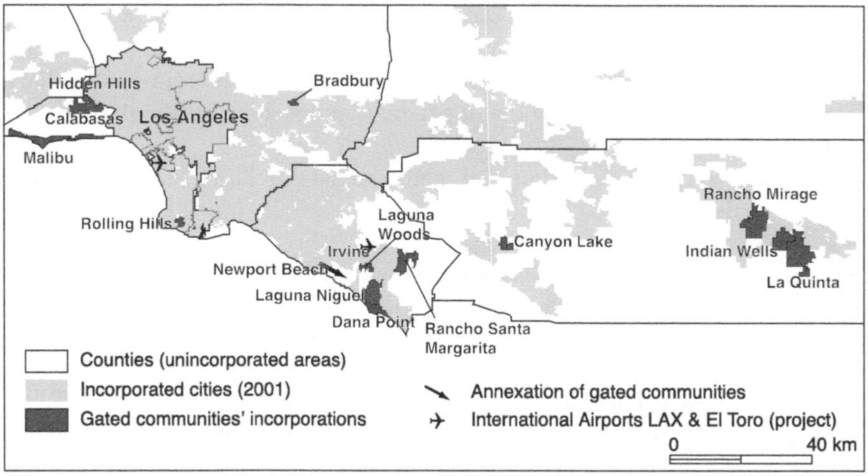

Figure 6.3 Gated communities and municipal incorporations in the Los Angeles area. Sources: US Bureau of Census Boundary Files, 2001, R. Le Goix – UMR Géographie-cités 8504 2002

of the use of tax dollars and maintaining a slow-growth policy favouring the leisure lifestyle and countryside setting (Yang and Ragland 1999). According to William O Talley, the city manager, incorporation was also a means to oppose the international airport project, west of the proposed city.[7] It is nevertheless unclear why Coto de Caza rejected incorporation, but it seems that wealthy Coto residents were concerned about Rancho Santa Margarita (including several middle-class neighbourhoods in addition to Dove Canyon; see figure 6.2) seeking to take control of Coto's high fiscal basis. Furthermore, it was reported that Coto is also planning to apply for incorporation on its own. It should be noted that Dove Canyon was highly motivated to join Rancho Santa Margarita, as the county offered the proposed city US$200,000 a year (financed from the county's share of property tax) in order to provide Dove Canyon with public services (Boucher and Gale 1998; Ragland 1999; Tessler and Reyes 1999). Private and gated Dove Canyon was confronted with an offer that met its own interests, entered into the deal and benefited from the more secure source of tax funds for service provision.

Setting aside the usual arguments of better local control of planning and the preservation of life style and identity, the incorporation of gated communities as new municipalities indeed reveals their political nature and has implications for other communities. Considering property values as a good proxy for assessing the tax base (given that property tax is the main resource for local governments), it appears that a majority of gated-community incorporations, when they took place, withdrew from the unincorporated areas a high-level tax base whereas nearby communities had to incorporate on a weaker fiscal basis (around Rolling Hills, Canyon Lake, Rancho Santa Margarita, Dana Point; see figure 6.2). In Los Angeles County, on the outskirts of the continuously urbanised area, low-end communities might even remain unincorporated because they have insufficient resources to incorporate, and represent a charge for a county constantly losing the upper-market developments built on its land (Miller 1981).

I further argue that, in doing so, gated enclaves seek to reach three interrelated goals:

- to prevent their up-scale fiscal basis from being redistributed in other (poorer) areas; this is a permanent and rather obvious goal in all incorporation driven by upper-scale developments in the US;
- to create a legal means of transferring public resources and assets for the profit of exclusive and enclosed neighbourhoods;
- and to find a legal means of getting public financing of infrastructure within gated areas, whereas such public infrastructures are generally incompatible with gated private enclaves.

The last two goals make gated communities predators of public resources.

Gated communities as predators of public resources

Through the incorporation cases previously exposed, not only do gated communities establish themselves as respectable public actors, but they also find public funding for private liabilities that were previously the responsibility of the POA.

An important impact of gated communities playing the game of municipal incorporation arises from their tendency to act as predators of public resources and to shift the use of city services (paid by the local taxpayers, along with other public grants) to the exclusive use of gated enclaves. No public money can be spent within the gates – otherwise public access would have to be granted and the gates would eventually become redundant. Such issues are demonstrated by the 1992 decision of Hidden Hills to build its city hall outside its gates in order to allow public access to the facility without opening the gates (Ciotti 1992; Stark 1998). The 1994 *Citizen's Against Gated Enclaves (CAGE) vs. Whitley Heights Civic Association* case banned the gating of public streets (Brower 1992; Kennedy 1995; McKenzie 1994). In 1999 Coto de Caza rejected a project to build a public school within its gates because it would have allowed the public inside the gated community (Nguyen 1999).

Such behaviour risks gated communities becoming prisoners of their own small worlds – with the risk of obsolescence if reinvestment funds are insufficient (Berding 1999). Approximately two decades after gated development started to occur, infrastructure is starting to wear out; streets need costly maintenance, costs are rising and it seems increasingly difficult to raise new funding. New special assessments or increased property owners' fees are problematic, especially because of the two-thirds majority approval required according to CID regulations (*Davis-Stirling Common Interest Development Act*, 1985). The issue can be sensitive in retirement communities where the average life expectancy of residents can be estimated at about five years (hence the comment already quoted in respect of the private governance quagmire in Leisure World – they 'don't buy green bananas'). In fact, many private non-gated CIDs can ultimately rely on the municipality to pay for the maintenance of streets and major infrastructure like parks and sewers. In Irvine many private neighbourhoods are maintained with public money, but streets are kept opened.

As a consequence, the only way to get public funding for a gated community is to rely on a public government to externalise services that were paid for by the POA and may be provided by a public entity without infringing the law, and without having to remove the gate. This can be done in several ways: a public service (library, kindergarten, trash collection, transit system and so on) that used to be in the charge of the POA can be transferred to the municipality; a contractual agreement can also authorise the police department to patrol within the gates (like in Dove Canyon), thus saving the POA some security and regulation enforcement costs.

A few examples further illustrate this assertion. First, incorporation allows the transfer to the municipality of the cost of services formerly paid for by the POA. In the peninsular community of Rolling Hills (630 housing units) all leisure facilities located inside the gates are the property of the city and the POA rents them with an exclusive privilege (Rolling Hills is the only gated community in the city and the municipality's jurisdiction fits the walls of the POA). Major maintenance costs are nevertheless borne by the City. In Calabasas, the developer of Parkway Calabasas gated communities created a community facility district[8] and contracted in 1992 a US$30-million loan to pay for beautification and building of parks and access roads to the gated communities. Because of its cost (US$4,000 a year by each resident), the loan was refinanced by the city in 2000 and this became the responsibility of all the taxpayers in Calabasas. Furthermore, 13 per cent of the municipal budget is dedicated to services paid for by the city, obviously favouring gated communities.[9] Although the golf course and other leisure facilities are supposed to be open to all the residents in the city, their location on the western side of the city, nested in the middle of gated enclaves, is detrimental to the open and middle-class neighbourhoods on the eastern side. Not only do the properties abutting the golf course in gated communities benefit from higher property prices, but also other residents and taxpayers do not enjoy the same level of services near their homes, and must undertake longer trips to access the public leisure facilities.

In the elderly gated community of Leisure World/Laguna Woods (both the city and the POA encompass the same area), besides the airport issue, the incorporation was also motivated because of rising costs, ageing private infrastructure and lower revenues from assessments than other gated enclaves. The retired population is indeed in a more modest condition than are residents on the eastern side of Orange County (figure 6.2). In response, sewer maintenance, trash collection and public transit are being transferred from the POA to the city. Incorporating Leisure World is also an answer to the intricate decision-making process in the POA previously explained: the municipality was designed to substitute for the POA when strategic and costly decisions are to be made, and is now in charge of zoning and planning. For instance, the city set up in 2001 the long-range planning requirements in order to forecast the improvements needed by the gated community's infrastructure.[10] The POA shall now comply with regulations enacted by a municipality that was first designed to be instrumental to the association. This exemplifies how a public authority may substitute for private urban governance when private management fails properly to govern a 19,500-inhabitant private enclave.

The recent annexation of gated communities on Pelican Hills (10,000 inhabitants forecasted by 2010) by the City of Newport Beach offers another dramatic example of gated communities seeking access to public funds. Because these communities are located under the landing and takeoff path of a planned airport, Newport Beach has proposed to annex them in order capture a new tax base (US$2.8 billion when the development will be completed by 2010), and also to prevent them from becoming dangerous opponents to the airport by incorporating later as a city in their own right. First, some residents challenged the annexation and the airport, but they did not manage to gather enough signatures on a petition to block the annexation. Second, once integrated among the 37,200 other residents of Newport Beach, the 7,000 residents (12,000 forecasted by 2010) do not represent a significant opposition force. And, last but not least, the motivation for not challenging the annexation has been guaranteed by the municipality of Newport Beach paying off an estimated US$18-million special assessment debt owed by gated-community residents, and offering the building of a US$7-million community centre, free trash collection and finally landscaping of areas outside gated communities (Willon 2001). This poignantly demonstrates the need for viewing gated communities as not only assemblages of private land, streets, infrastructure and services but also as political forces that will make decisions and play games in their members' interests – including preying on public resources.

Conclusion

The novelty of gated communities in the suburban landscape is their enclosure and its implication for the nature of the territory developed behind the gate. Theoretically speaking, gated communities are private areas entitled to provide public services privately. The sprawl of gated communities is not to be understood as secession from public authority, but as a public–private partnership: the gated community provides the public authority with new taxpayers at little cost, while the property owners' association is granted autonomous local governance and assumes responsibility for local urban affairs. Nevertheless, gated communities tend also to become public actors (municipalities) and can be expected to try and offset the burden of private governance by transferring costs to the municipal entity, using public funds and federal grants where possible for the exclusive use of private enclaves. Municipal incorporations have helped gated communities to acquire the status of real public actors, providing the private interest with official representation on the public scene.

This peculiar use of a public local government produces a cost borne by the urban community as a whole because of the scope for bargaining and other game-playing. This adds another dimension to the spillover effects of gated communities that have been widely discussed, including the impact on property values outside gated communities and the increase of segregation patterns (Le Goix 2003a, 2003b).

Notes

1 This paper is drawn from a doctoral thesis funded by the CNRS (UMR Géographie-cités 8504, Paris), the French–American Foundation (Tocqueville Fellowship, 2000–1), and the French–American Commission (Fulbright Research Scholarship, 2002–3), which is gratefully acknowledged.

2 Because of the lack of a comprehensive survey of gated communities at a local scale, this research is based on a database derived from the sources a prospective home-buyer would use. Once integrated within a geographical information system with 2000 Census data, the diversity of the market can be assessed, as well as the location of gated communities and their social patterns. Accompanied by field surveys, and by interviews with gated communities and local officials, the most relevant sources for locating gated neighbourhoods were *Thomas Guides®* maps plotting gates and private roads, real-estate advertisements in the press and in real-estate guides, and county assessors' maps. A total of 219 gated communities built before 2000 have thus been identified in seven counties (Los Angeles, Riverside, Orange, Ventura, San Bernardino, Santa Barbara and San Diego).

3 According to the *2001 American Housing Survey*, it can be estimated that on average 11.7 per cent of the households are in walled, fenced and access-controlled communities in the Los Angeles Metropolitan Area, based on a national sampling of households (Sanchez, Lang and Dhavale 2003). It is relevant to mention here that the research presented in this chapter relies on a more restrictive sampling of gated communities, designed to exclude the condominiums and secured apartment complexes, which do not include privatised public spaces, according to Blakely and Snyder's definition of gated communities (1997). The analysis excludes any vertical co-ops or condominiums in which common areas are limited to parking, a common garden or a swimming pool.

4 Interviews with Kathy Bennet, City of Canyon Lake Clerk, and Linda Musselwhite, Canyon Lake POA Member Service Manager, December 2000 (City of Canyon Lake 2000).

5 Interview with Agnes Flore, Finance Office, November 2001, and the municipal booklet *A Look Back in Time* (City of Rancho Mirage 1993).

6 Interview, November 2001.

7 Interview, November 2001.

8 A community facility district (CFD) provides public services (improvements, water adduction and so on) to a designated area. According to the 1982 Mello-Roos Act,

CFDs are usually set up by developers and are financed by bonds and special assessments paid by homeowners included within the CFD boundaries.

9 Interview with Donald Duckworth, City Manager of Calabasas, December 2001 (City of Calabasas 2001).

10 Interviews with R. Ring and Leslie Kean, Laguna Woods City Manager, December 2001.

Bibliography

Berding, T.P. (1999) 'The uncertain future of common interest developments', *Echojournal: A Journal for Community Association Leaders*, 5: 5.

Bjarnason, S.J. (2000) 'Lawn and order: Gated communities and social interaction in Dana Point', unpublished thesis, University of Oregon.

Blakely, E.J. and Snyder, M.G. (1997) *Fortress America: Gated Communities in the United States*, Washington, DC and Cambridge, MA: Brookings Institution Press and Lincoln Institute of Land Policy.

Boucher, G. and Gale, E. (1998) '2 S. County areas are walls – and worlds – apart', *Los Angeles Times*, 9 August: 1

Brower, T. (1992) 'Communities within the community: Consent, constitutionalism, and other failures of legal theory in residential associations', *Land Use and Environmental Law Journal*, 7 (2): 203–73.

Caldeira, T. (2000) *City of Walls: Crime, Segregation, and Citizenship in São Paulo*, Berkeley, CA: University of California Press.

Ciotti, P. (1992) 'Forbidden city', *Los Angeles Times*, 9 February: B3.

City of Calabasas (2001) *Community Facility District no. 2001–1, Special Tax Refunding Bonds, Financing Summary Report*, Calabasas, CA.

City of Canyon Lake (2000) *Proposed Operation and Capital Improvement Budget, Fiscal Year 2000–2001*, Canyon Lake, CA.

City of Rancho Mirage (1993) *A Look Back in Time*, Rancho Mirage, CA.

Davis, M. (1990) *City of Quartz: Excavating the Future of Los Angeles*, London: Verso.

—— (1998) *Ecology of Fear: Los Angeles and the Imagination of Disaster*, New York: H. Holt.

Dear, M. and Flusty, S. (1998) 'Postmodern urbanism', *Annals of the Association of American Geographers*, 88: 50–72.

Donzelot, J. and Mongin, O. (1999) 'De la Question sociale à la question urbaine', *Esprit*, 258: 83–6.

Flusty, S. (1994) *Building Paranoia: The Proliferation of Interdictory Space and the Erosion of Spatial Justice*, West Hollywood, CA: Los Angeles Forum for Architecture and Urban Design.

Foldvary, F. (1994) *Public Goods and Private Communities: The Market Provision of Social Services*, Aldershot: Edward Elgar.

Frey, W.H. (1993) 'The new urban revival in the United States', *Urban Studies*, 30 (4/5): 741–74.

Jaillet, M.-C. (1999) 'Peut-on parler de sécession urbaine à propos des villes européennes?' *Esprit*, 11 (258): 145–67.

Kennedy, D.J. (1995) 'Residential associations as state actors: Regulating the impact of gated communities on nonmembers', *Yale Law Journal*, 105 (3): 761–93.

Lacour-Little, M. and Malpezzi, S. (2001) 'Gated communities and property values',

Wisconsin-Madison CULER working papers, 01 (12): 1–36. Online. Available HTTP: http://ideas.repec.org/p/wop/wisule/01-12.html (accessed 17 August 2004).

Le Goix, R. (2002) 'Les gated communities en Californie du Sud, un produit immobilier pas tout à fait comme les autres', *L'Espace géographique*, 31 (4): 328–44.

—— (2003a) 'Les "Gated Communities" aux Etats-Unis. Morceaux de villes ou territoires à part entière? [Gated communities within the city in the US: Urban neighbourhoods, or territories apart?]', unpublished thesis, Université Paris 1 Panthéon – Sorbonne. Online. Available HTTP: http://tel.ccsd.cnrs.fr/documents/archives0/00/00/41/41/index_fr.html.

—— (2003b) 'Gated communities sprawl in Southern California and social segregation', paper presented at international conference on gated communities: Building Social Division or Safer Communities, Glasgow, September 2003. Online. Available HTTP: http://www.neighbourhoodcentre.org.uk/gated.html.

McKenzie, E. (1994) *Privatopia: Homeowner Associations and the Rise of Residential Private Government*, New Haven, CT: Yale University Press.

Marcuse, P. (1997) 'The ghetto of exclusion and the fortified enclave: New patterns in the United States', *American Behavioral Scientist*, 41: 311–26.

Miller, G.J. (1981) *Cities by Contract*, Cambridge, MA: The MIT Press.

Nguyen, T. (1999) 'Coto de Caza residents say no to school within gates', *Los Angeles Times*, 4 March: 1.

Purcell, M. (1997) 'Ruling Los Angeles: Neighbourhood movements, urban regimes, and the production of space in southern California', *Urban Geography*, 18 (8): 684–704.

Ragland, J. (1999) 'City-to-be, county agree on tax deal; Cityhood: Rancho Santa Margarita will get $200,000 per year; boundaries to include Dove Canyon, Robinson Ranch', *Los Angeles Times*, 23 June: 5.

Reich, R.B. (1991) 'Secession of the successful', *New York Times Magazine* 20 January: 16.

Sanchez, T., Lang, R.E. and Dhavale, D. (2003) *Security versus Status? A First Look at the Census's Gated Communities Data*, Alexandria, VA: Metropolitan Institute, Virginia Tech.

Sorkin, M. (1992) *Variations on a Theme Park: The New American City and the End of Public Space*, New York: Hill and Wang.

Stark, A. (1998) 'America, the gated? (impact of gated communities in political life)', *Wilson Quaterly*, 22 (1): 50–8.

Tessler, R. and Reyes, D. (1999) '2. O.C. Gated communities are latest to seek cityhood', *Los Angeles Times*, 25 January: 1.

Webster, C., Glasze, G. and Frantz, K. (2002) 'The global spread of gated communities', *Environment and Planning B: Planning and Design*, 29 (3): 315–20.

Willon, P. (2001) 'Petition to stop Newport Coast annexation fails', *Los Angeles Times*, 22 November: 3.

Yang, E. and Ragland, J. (1999) 'New Year likely to bring O.C. new city', *Los Angeles Times*, 24 June: 1.

7

CONDOMINIOS FECHADOS AND *BARRIOS PRIVADOS*

The rise of private residential neighbourhoods in Latin America

Michael Janoschka and Axel Borsdorf

Abstract

Private and gate-guarded residential neighbourhoods have sprawled in all major Latin American city regions during the last two decades. The aim of this text is an analysis of the spatial and social implications of the 'condominisation' of the urban landscape within a theoretical and empirical frame. Starting from a historical background, the actual patterns and consequences associated with the distribution of gated communities are discussed and presented in a model of the current Latin American city structure. Further attention will be focused on the relation between urban insecurity and the spread of the private neighbourhoods based on empirical data gathered via narrative and biographical interviews with inhabitants. From our results we conclude that insecurity and crime rates are not the driving forces motivating life behind gates but rather the achievement of an ideal lifestyle.

Introduction

Buenos Aires' inhabitants proudly introduce their city to visitors by enumerating records. Taking a taxi from the international airport to the city centre, everyone will learn that the Plate River is the broadest river crossing an agglomeration in the southern hemisphere or that 9th of July Avenue is the widest urban avenue crossing the densest central business district (CBD) in South America. By the late 1990s the Argentinian capital had notched up two more records: the widest metropolitan motorway on the subcontinent (16 lanes) links the dense city centre with Nordelta, Latin America's largest gated development. As we shall indicate, the so-called town-village (*CiudadPueblo*) of Nordelta contrasts with the rest of the 14-million-inhabitant mega-city, in which approximately 60 per cent of the population live on a daily income less than one euro.

> An avenue surrounded by palms. Great green areas and parks. Lime trees, spruces, willow-trees and magnolias. An enormous and silent water surface. And everything that design and comfort can nowadays introduce to achieve a better life. A place like this does exist. And it is not at the end of the world. It is exactly in the geographical centre of Nordelta. Its name is the Island.
>
> (Nordelta 2002)

Although Nordelta is far from being an isolated island, the marketing target is based on selling an image of a grand, self-sufficient garden city completely detached from the local and national economic and social setting. The 1,600-hectare development for about 80,000 inhabitants contains, among other facilities, a huge sports complex, its own private schools, a technical school and a private university. Despite the proximity to commercial and leisure facilities in nearby suburbs, Nordelta has its own mall and urban entertainment centre and also a Civic Centre (figure 7.1). These spaces have a pseudo-public character. Common spaces are available to Nordelta residents and help create a sense of local citizenship and social interactions, which is otherwise inhibited by the access controls at each of the 30 or so planned neighbourhoods inside Nordelta (eight of which are functional as of 2001). Public life in this sprawling urban club is well controlled, however, distinguishing it from equivalent spaces outside. Surveillance cameras monitor the use of public space and the unusual has a strong chance of being detected. Every point of the development can be reached by security or service vehicles within a 90-second response time.

Analysing the wish to attain distinction and the citizenship-building processes occurring in Nordelta, it is easy to conclude that this is not only an official strategy of the real-estate companies to improve the sale of the properties. Individuals living inside also follow a personal strategy that stresses the change in personal lifestyle that took place after moving to Nordelta. Many inhabitants refer to themselves as *Nordelteños*, in contrast to the local characterisation of Buenos Aires' inhabitants as *Porteños* (Janoschka 2002a: 90). Education institutions pick up the same nuance, such as the Catholic Cardinal Pironio College's promise of a 'new education for a new civilization'.

Nordelta is still the biggest and most complex private urban development in Latin America, but it is not alone. With the exception of Cuba, gated and access-restricted neighbourhoods have become a key part of the real-estate market throughout the whole continent. In this chapter, we comment on the historical evolution and contemporary importance of gated communities for Latin American city regions. The discussion focuses on the reasons for the proliferation of access-restricted neighbourhoods and the controversial question concerning urban insecurity. A comparative analysis based on empirical evidence from the authors' case studies in different cities and countries is used to characterise common aspects and differences in the phenomenon. We use a lifestyle-based analysis derived from our empirical work with inhabitants of gated communities. The likely political consequences for local administration

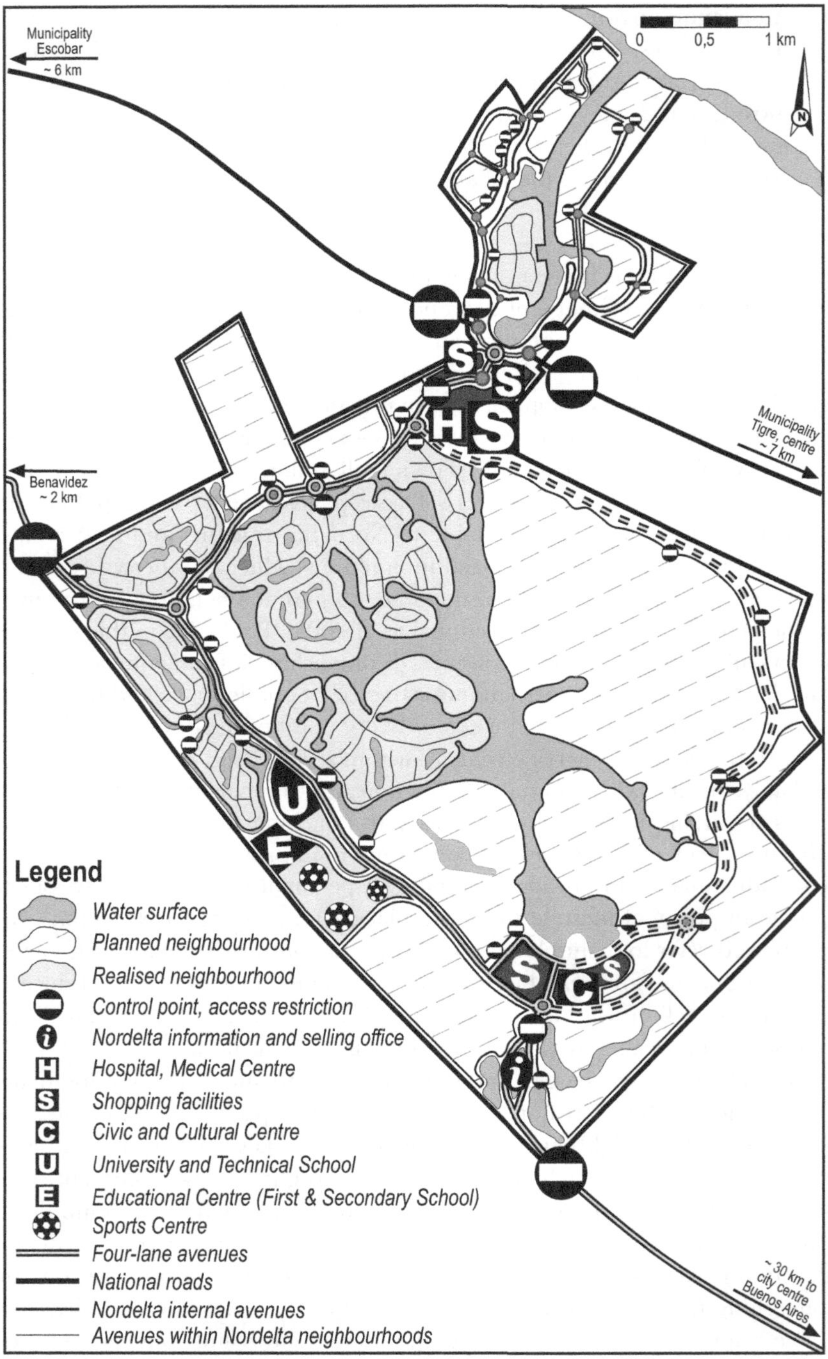

Figure 7.1 Town-village Nordelta, the major gated development in Latin America. Source: based on Janoschka 2002a, Janoschka 2004 and Nordelta 2002

and social contract between different classes are addressed in the conclusion. We begin with a short introduction to the socio-political and urban transformation processes in Latin America.

From polarised city to fragmented agglomeration

Massive changes have been taking place in Latin American societies since the early 1980s, when re-democratisation processes started to replace the military regimes of the previous decade. Regardless of the ideological orientation of the new democratic governments, capitalist and neo-liberal politics were implemented. That time marked the end of the development strategies employed since the start of the post-war period, strategies that were based on the substitution of imports by domestic industrialisation. This industrialisation was accompanied by import restrictions and active economic intervention by the public sector. From the 1980s this development strategy began to change towards one of integration into world markets, with subsequent lowering of tax barriers. Privatisation of state-owned companies and former state-organised services and a new pro-market ethos led to a complete reorientation of Latin American economic activity. Foreign direct investment and subjugation to open market laws led to a reduction of the importance of the public sector and in many countries induced a deindustrialisation process due to the greater competitiveness of imported products. In the social sphere, the results of the neo-liberal economic trend included a substantial increase in social polarisation (cf. Ciccolella 1999) and the rise of unemployment, despite high economic growth rates experienced by most Latin American countries in the 1990s.

Following the finance market crisis of the late 1990s, Latin America's economic panorama notoriously changed. With the exception of Chile and Mexico, the latter country profiting from NAFTA integration with the USA and Canada, the dynamics of Latin American economies became negative. Reverse capital flows were experienced from 1998 and still continue. Economic growth has been low and the income per capita has been slowly decreasing. Some countries, such as Argentina, Uruguay or Venezuela, suffered strong economic depressions accompanied by internal political problems. Under such conditions, social polarisation has accelerated (Ocampo 2003; Ruiz 2003).

In South American cities over the last 15 years, there has been a rapid modernisation of the urban infrastructure, driven in many cases by foreign investment. This investment was concentrated partly on basic urban services such as telephone and water provision, which are now owned by international companies providing an international standard of service for the middle and upper classes. However, investment also took place in the trappings of modern globalised cities: tolled motorways, private industrial parks, international hotel chains with integrated business facilities, shopping malls and hypermarkets, urban entertainment centres, multiplex cinemas and gated and access-restricted residential quarters in the city and the suburbs (Janoschka 2002b). The absence of state intervention in urban planning led to most new private investment being directed towards an exclusive urban form, based on private transport. A new car-based lifestyle flourished, encouraging fragmentation and spatial segregation. While earlier decades of urban growth were driven primarily by lower-income migration from rural areas, the pattern has changed since the 1980s, with migration strongly declining. In the last two decades urban expansion has been substantially a result of the greater use of space per capita, chiefly driven by changes in residential areas of the upper and middle classes.

A model of fragmented agglomeration

During the period of development strategies led by import reduction, ending in the 1980s, one of the distinctive features of Latin American cities was the strong polarisation of urban spaces. Differentiation in social status was heavily bound to location within the city. Urban space was divided into polarised sectors creating a strongly differentiated rich city and poor city (Gilbert 1998). This has changed during the last two decades. Settlements of poor and rich people moved closer to one another due to the occupation of suburban and formerly poor areas by high-income populations. Poor families also installed themselves in abandoned, formerly homogeneous, middle- and upper-class districts. This process has been accompanied by a stronger delimitation of small areas (micro-neighbourhoods) often accompanied by private security services. This helps explain the function of gated communities in Latin American cities. They are homogeneous, highly segregated and protected areas allowing the middle and upper classes to cohabit increasingly scarce space. They are reshaping the physical and social ecology of urban space (Sabatini, Cáceres and Cerda 2001; Parnreiter 2004; Borsdorf and Hildalgo 2005; see figure 7.2). They have become the new model of Latin American urban agglomeration.

In 1976 German geographers Bähr and Borsdorf published two independent models of Latin American city structure which were the starting point for an intense discussion in Europe and North America (Griffin and Ford 1980; Gormsen 1981; Deler 1992; Crowley 1995; Bähr and Mertins 1981; Borsdorf 1982). Following this, the dramatic transformations which occurred in Latin American city regions from the 1980s, were related to new theoretical insights in the work of Ford (1996), Meyer and Bähr (2001) and Borsdorf (2002a), who redesigned the traditional models or developed new models (Janoschka 2002b, 2002c), derived from new empirical data. Borsdorf, Bähr and Janoschka (2002) suggested a model which integrated the different arguments of the other authors (figure 7.3). City development in Latin America is characterised in four phases, which correspond to different modes of state intervention (urban planning) in the real-estate market.

In the early phase, the colonial time, the dominant planning principle was compactness of the urban body and a social gradient according to distance from the central plaza. Social status was higher close to the plaza due to the central social and political functions concentrated there and the high cost and the time of travel under available technology. This pattern gained importance

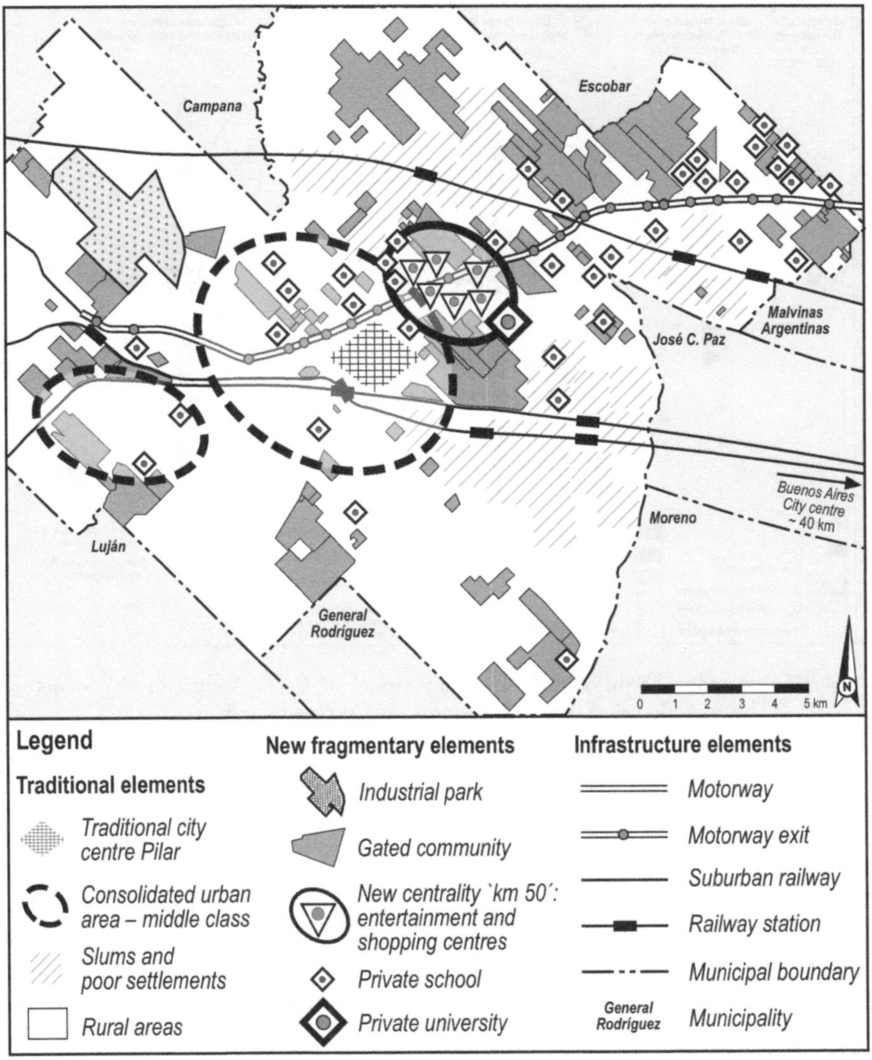

Figure 7.2 Fragmented urban space in the suburban municipality of Pilar in Buenos Aires, 2004. Source: Janoschka 2004

due to royal instructions for the foundation of cities, the subsequent location of the higher-order urban functions in and around the central plaza and the settlement of the colonial elite around the plaza.

After independence, linear structures began to gain in importance. The upper classes moved to new houses extending along the main street – the *prado*. The most important streets then became the location for market gardening and artisan activities and, later on, early industrial activity. These sectoral patterns did not completely change the older ring structure, but strongly transformed the urban format.

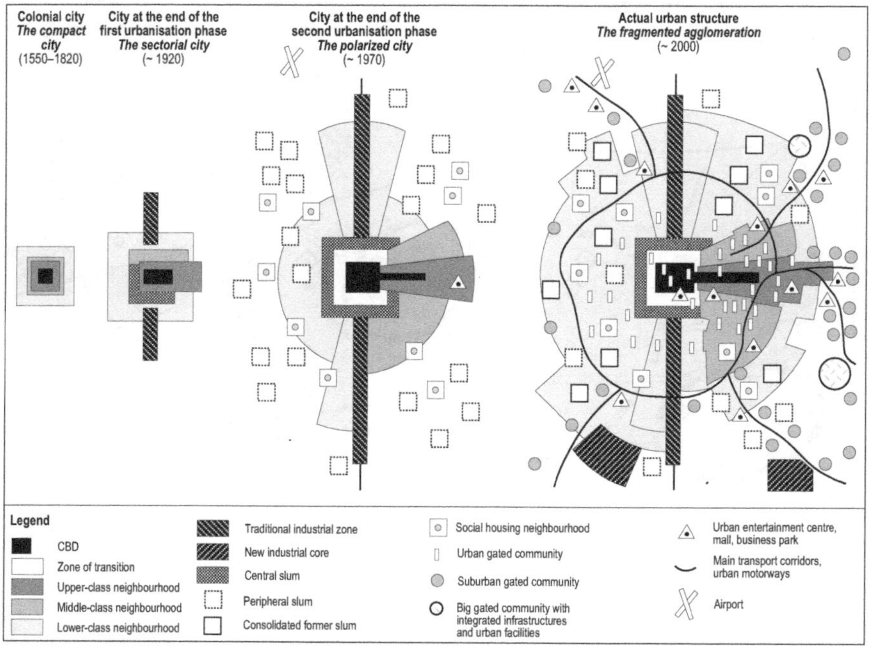

Figure 7.3 Borsdorf, Bähr and Janoschka's model of Latin American city structure. Source: Borsdorf, Bähr and Janoschka 2002; adapted

From the 1930s onwards movement from rural areas to urban centres caused a massive population increase, and the agglomeration process began to display intense polarisation between rich and poor areas, implying a development pattern typical of the modern industrial city (Marcuse 1989). The upper-class areas transferred from the centre towards the suburbs, but the main feature was the expansion of poor neighbourhoods. Formal urban political institutions became less and less able to set the guidelines for urban development, not only due to high growth rates, but also because of corrupt and polemic policies which alternated with authoritarian regimes during phases of dictatorship. The sectoral principle of urban growth remained broadly intact but was overlaid by the polarisation principle. Only some time later in this period did a process of cell-type growth begin in the peripheries. In suburban or outer urban areas there began to emerge neighbourhoods of low-income housing such those in illegal marginal areas (*favelas*, *villas miserias*), creating a cellular pattern, which started to gain importance.

During the last two decades, as we have indicated, urban expansion took a different direction. To be more specific, contemporary development can be characterised along the following lines (Janoschka 2002c: 65f):

- Gated communities for the upper and middle classes are appearing throughout the whole metropolitan area, breaking with the sectoral concentration of upper-class areas.
- The dispersion of malls, shopping centres and urban entertainment centres is taking place across the whole agglomeration, no longer concentrated in the upper-class sectors.
- Gated communities occupying a larger area and integrating more and more urban functions are becoming more common. Complexity and size of major gated communities are passing the level of small cities, particularly in São Paulo and Buenos Aires, but also in smaller cities such as Córdoba in Argentina.
- New transport infrastructure is helping shape this dispersal, with proximity to a motorway entrance being the most important influence shaping the spatial pattern of demand for detached housing areas.
- Industrial production is becoming suburbanised and logistical activities located in peripheral industrial or business parks.
- Lower social groups and marginal areas are increasingly segregated through walls or more informal barriers to access.

The crucial spatial influence on the new fragmentary and nodal urban structure is the transformation of the transport system, especially the motorways. These have been transformed from insufficient and saturated low-capacity highways to modern and effective ones. This has chiefly been made possible by private investment and so the stimulus for widespread private-led urban development under the neo-liberal political and economic governance regimes has been private-led modernisation of the transport infrastructure. Strong reduction in commuting times made suburban locations newly interesting to the middle and upper classes, whose valuation of time is that much greater.

As a result of the increased accessibility of the urban periphery, the cellular developments there, which in former times were predominantly in marginal areas, are becoming more integrated into the market sphere and increasingly interesting investment areas for real-estate enterprises. Another feature of the fragmentation process relates to the location of different urban functions. This is especially notable in the retail sector. Although the urban centres have gained new importance due to renovation programmes and other upgrading intervention, they cannot compete with suburban malls, which represent a North American lifestyle of rising popularity. Social exclusion works here not through the construction of walls but by accessibility. These islands of consumption and leisure are generally accessible only to those who have access to a car. However, while the first malls were clearly oriented to the upper-class market in the urban upper-class sector, such facilities are now found in the whole urban area due to the dispersion of the upper and middle class. This implies a degree of improvement in the access of the poor. Poorer areas also have been transformed during the last two decades. Marginal areas have been integrated into the city and in some cases have been upgraded. On the other hand, there are numerous cases of marginal areas which have resisted any integration and are now almost inaccessible nodes of criminality.

This discussion of the transformation of Latin American city structure shows that urban enclosure of the environment and the resulting fragmentation of accessibility is not only the preference of the high-income population, but a more general societal principle. In order to understand the specific case of Latin America better, a historical analysis of the rise and diffusion of gated residential neighbourhoods is presented in the next section.

The history of gated residential neighbourhoods in Latin America

Discussion about gated communities in Latin America is dominated by characterisation of the new aspects of urban dynamics (Meyer and Bähr 2001; Coy and Pöhler 2002), the growing importance of a North American-dominated lifestyle and the impacts of global real-estate firms and a globalised elite labour force and consumption sector. Globalisation and the fear of criminality are two central arguments used to explain the proliferation of gated communities. These aspects should not be neglected, but analysing the appearance of gates from a historical point of view tempers the macro-explanations with local specificity. It is possible to find important lines of explanation in the typical behaviour of the Latin American elites through decades if not centuries.

The Spanish colonial cities of Latin America were open and without city walls. But this openness was accompanied by strong internal closure. The residential buildings – atrium houses with internal patios – were closed towards the street with heavy wooden doors, and until the mid-19th century windows were small and located to guarantee privacy. Even inside the buildings, different grades of segregation and privacy were established; guests never had access to the second or third patio, which was reserved for women or for garden activities (cf. Borsdorf 2002b). Also in the colonial epoch, enclosed neighbourhoods for multiple households, notably widows, orphans or priests, are reported in cities such as Santa Catalina in Arequipa or the Hospitolio Cabañas in Guadalajara (cf. Borsdorf 2002b: 239). During the 19th century, another form of access-restricted neighbourhood emerged – company towns for the workers and employees of mining companies. These were highly segregated and in many cases walled or with other public-access restrictions. These examples show that the principle of gating has occurred in Latin American cities throughout history, but this does not differ so much from what is known about European cities.

The more significant precursor to the modern closed neighbourhood was the presence of poor households in the atrium houses of central areas. This occurred from the late 19th century, when the owners of such houses moved towards the latest fashionable areas of the expanding city. Single-family atrium houses were subdivided, and each room rented to a low-income family. This transformation occurred in all bigger Latin American cities. The housing areas were called *tugurios* in Peru or *vecindades* in Mexico. In Chile and Argentina the term *conventillo* semantically connotes the concept of the walled sacred communities already referred to. The wooden entry gates of the former patio houses were transformed into 24-hour, guarded entrances to prevent the entry of new neighbours who did not pay rent. When the historical building stock was filled, this kind of structure was also reproduced in new lower-income areas, for example in Chile during the 1920s and 1930s or in Mexico until the early 1950s. Hidalgo (1999) and Rovira (2002) point out the similarity of these neighbourhoods to the gated communities of lower-income groups in the late 20th century. In a context of political instability, military regimes and the loss of state control, the gated communities of these groups become comprehensible presently and historically.

A parallel exists with many Latin American slum areas on city outskirts, called *favelas* in Brazil and *villas miserias* in Argentina. In most cases, the origins of these settlements, which sprawled in Latin America from the 1950s, are irregular if not outright illegal. The lack of legality led to a series of hardline state interventions including mass evacuations and demolition in the 1960s and 1970s. This followed and exacerbated a rising criminalisation of these areas (headquarters of drug dealing, organised criminality) and it induced the creation of internal organisation structures to prevent the access of persons living outside. These were not created in the same way as modern entrepreneurial gated communities but the effect is very similar. The public cannot enter and a part of the city has become enclosed and locally governed.

There are also historical precursors to gated communities for the upper classes. Some authors have suggested that the origin of the 'wealthy' gated communities can be traced to the country clubs or golf club model. Borsdorf (2002b) describes the case of Mexico City, where, at the beginning of the 20th century, golf was imported by wealthy English immigrants and became rapidly popular among the European upper class. In the case of Buenos Aires, the first country clubs were built in the early 1930s, also emulating the lifestyle of the English-oriented bourgeoisie (Janoschka 2002d). At the time of their foundation, such clubs were still far away from the city, but the rapid growth of the agglomeration brought them within the urban core during the second half of the 20th century. Clubs found it necessary to guard their territories and secure the entrance due to low occupation during the week and the close proximity of potential trespassers. Up to the early 1970s the number of these clubs was low and they were still mostly used for vacation and weekend activities. From that time, however, a first wave of diffusion of such clubs was observable. As an example, about 70 country clubs for the upper and upper middle classes were founded in Buenos Aires in the mid-1980s, mainly for sports activities in a quasi-natural environment.

Gated neighbourhoods in the last two decades

Linking today's wealthy gated communities with their historical antecedents, advertisement remains focused on connecting gated housing with an elitist lifestyle, as the quotation from a Nordelta publicity leaflet shows clearly at the beginning of this chapter. Until the mid-20th century, European elites provided the reference points; now it is North America. Many advertisements, especially in Colombia and Brazil, offer a recipe for achieving this target lifestyle. Photos and remarks about lifestyle in Miami accompany the discourse of fear and insecurity which is especially strong in those two countries. The result is the construction of exotic habitats, including a range of temperate-zone plant species not naturally occurring in Latin America.

Constructing a typology of these developments begins with a problem of definition. Throughout Latin America, a range of terms are in common use. The meaning of the same word may substantially vary in different countries. *Condomínio fechado* (Brazil), *barrio privado* (Argentina) or *fraccionamiento cerrado* (Mexico) stand for different products of the real-estate market in the different countries, and even the term *barrio cerrado* (closed neighbourhood), used in various countries such as Chile, Ecuador and Bolivia, has a wide range of meanings. Therefore we adopt a consistent language using English terms.

Three main types of gated neighbourhood can be differentiated by structure, location and size. The three groups are defined as follows:

- Urban gated communities, which can be defined as groups of attached houses or tower blocks with few common facilities. The target of these developments may be middle- or lower middle-class families in intermediate locations (even social housing projects) or upper middle- to upper-class families in central areas where land shortage limits the development of common facilities. This category also includes the enclosure of existing areas, in most cases upper-class single-family housing areas in central or intermediate locations.
- Suburban gated communities, predominantly for the middle and upper classes with single detached houses. These developments share a wide range of common facilities including sports and community facilities and landscaped gardens. There are also suburban gated communities without common facilities and these may be oriented to lower middle-income groups.
- Mega-projects with integrated cultural and educational facilities. Although there are still only a few of these projects, the dynamics and the internationalisation of the real-estate market make this the most rapidly growing segment.

All three types have boomed throughout Latin America. Between 1990 and 2001, in Mexico City, around 750 new gated neighbourhoods with almost 50,000 housing units were launched on the market (Parnreiter 2004). In the Argentinian capital Buenos Aires more than 450 suburban gated communities currently exist, 80 per cent of them were started between 1995 and 2001. A dozen of these reach a size of more than 5,000 inhabitants (Janoschka 2002a). To this figure an unknown number of urban gated communities known as garden-towers *(torre jardín*, cf. Welch 2002*)* must be added in Argentina. In 2002 more than 130 of these garden-towers were listed in the local newspapers as offering new apartments. The quantity of pre-existing stock is unknown. Estimation of the total population ranges between 300,000 and 600,000 persons. In another example, Pöhler (1999) estimates that more than 100,000 people live in *condomínios fechados* in Rio de Janeiro's upper-class and beach-oriented city expansion area called Barra da Tijuca. Brazil's financial centre and mega-city São Paulo also has a mega-project, Alphaville, with around 35,000 inhabitants and more than 100,000 people working inside the gates. Gated communities have also diffused to secondary cities such as Córdoba, Argentina (1.3 million inhabitants). Within the metropolitan area there are more then 50 gated residential developments; one of them, Valle Escondido (Hidden Valley), promoted with the slogan 'the new city', offers a cluster of different gated communities for a total population of approximately 25,000 inhabitants (Roca 2001; ECIPSA 2003). Data for Curitiba in Brazil estimate the number of gated communities at about 300. This pattern is repeated in most big Latin American cities. In medium-size cities (less than 500,000 inhabitants) and in some cases even in small towns (for example Gualeguaychú, Entre Rios, Argentina), gated communities are also becoming more and more common.

The market share of gated housing complexes is extremely difficult to estimate. There are poor statistical data for the construction sector and in most cases differentiation between the products is not clearly specified. For the case of Buenos Aires, our own calculation is that there are about 100,000 units in the city's suburban gated communities. In relation to the whole housing stock, which is about 3 million, the market share represents not more than 3 per cent. But considering only the proportion of suburban housing, the market share has already risen to 10 per cent. If the calculation is based on the demand group (the upper 15–20 per cent of the population in terms of income, equivalent to 450,000 households), these 100,000 units represent between 20 and 25 per cent of the market share. The market for single detached houses or lots for construction of detached houses in suburban locations is more or less synonymous with gated communities. During the period of macro-economic stability from 1991 to 2001, between 80 and 90 per cent of urban expansion was related to the expansion of gated communities.

To these figures should be added the 'quasi' gated communities; streets and neighbourhoods with strong vigilance and closure of access during the night or retrospectively enclosed neighbourhoods, which do not appear as gated developments in the open real-estate market. Also, social housing projects from the 1970s, which are nowadays walled and gate-guarded, add to the contemporary stock of enclosed neighbourhoods. The same happens with a whole range of areas which are inaccessible due to internal governance and the predominance of criminal structures.

Urban insecurity

Our earlier discussion showed that the proliferation of gated communities started in Latin America long before political decisions led to internationalisation of the region's economy. Neither globalisation nor Americanisation of society can adequately on their own explain the peculiar rise in gated housing areas in Latin American cities.

Explanations focussing on security seem at first sight rather convincing. Superficially there seems to be a temporal correlation between the socio-economic transformation of most Latin American societies, which led to a substantial rise of urban insecurity, and the proliferation of enclosed neighbourhoods (Caldeira 2000; Dammert 2001). The development of gated housing complexes seems to be a rational solution not only for the richest parts of the population, but also for everyone else. It is an obvious consequence of rising criminality. Empirical data from several studies in Latin America show, however, that concern about criminality cannot fully explain why people move to a gated neighbourhood. The exception is Colombia, where about 40,000 murders per year are committed due to haphazard and organised violence and internal political conflict. However, in all other cases there are more important reasons than violence in itself. Between different countries and cities, the significance of the fear of violence varies. Caldeira (2000) showed that for the Brazilian city of São Paulo, urban violence reached such high levels that most inhabitants of gated communities had direct experience of some kind of criminality before making the decision to move inside the gates. This experience is also found in several big Brazilian cities and also in Mexico City. But this may not be a representative picture for the rest of the cities in the hemisphere. Biographical interviews with inhabitants of Nordelta in Buenos Aires, for example, show that despite media discourse and marketing emphasis, which stress fear of crime, inhabitants of the private city do not mention these factors as key influences on their decision to move there (Janoschka 2002a). There are several aspects which are more important for the new inhabitants of Nordelta. These include:

- Political and economic insecurity in the wider city, accompanied by the incapacity of the state to organise urban services and social infrastructure.
- The enhanced urban and suburban environment in the gated city, created by high levels of private investment. On the one hand, suburban spaces were viewed as more attractive because of the better standard of infrastructure (motorways, private schools, shopping and entertainment facilities). On the other hand, the urban environment outside was viewed as being degraded due to increasing density, rising motorisation and state retreat from maintenance of public spaces.
- The desire for a change of personal lifestyle and the search for a more socially homogeneous environment. Nordelta inhabitants are mainly between 35 and 40 years old and have one or more children. The location

gives them space and time with the family and outdoor activities for the children. The decision to move into the private city is an option to change lifestyle and guarantee a secure and predictable quality of life for the family. The package purchased is not only the immediate benefits within the gates but also the chance of a high standard of education and future networks for the children, to help secure the long-term social standard of the family.
- Last, but not least, the wish to achieve a new lifestyle is motivated by face-to-face propaganda, personal knowledge and group behaviour. Most people moving to Nordelta know from relatives and friends how life is organised in the gated neighbourhood. Some people even commented on friend-group pressure as a determinant for their decision. In certain social groups you are 'out' if you do not move to the suburban gated neighbourhood.

Empirical data collected by the authors from other Latin American cities, such as Santiago de Chile, Mexico City or Quito, show similar patterns (Janoschka 2002e). Despite the huge differences in urban settings, social behaviour and socio-economic development between different countries in Latin America, the importance of criminality is much lower than expected and in many cases irrelevant.

Conclusion

The physical fragmentation processes analysed in this paper are having an impact on the nature of urban life. Inhabitants of Latin American gated communities rapidly change their lifestyle and fully adapt their daily habits to life within an access-restricted area. Local public spaces lose their basic role as points of interaction between different classes as each class organises its own homogeneous space 'inside'. This tendency rises with the growing complexity of urban functions within the new private cities, as shown in the example of Nordelta. Inhabitants are increasingly living live in bubbles which are detached from the surrounding local political and social environments. The central difference compared to former times is the fact that today all social classes show signs of wanting to escape from the wider public sphere.

Latin American elites never strongly oriented themselves towards their own society but followed lifestyles imported from Europe and later from the US. A life behind walls may lead gated neighbourhood inhabitants even further away from the social reality of their 'home' society, which for the great majority of the population consists of a struggle for daily survival. From this point of view, the proliferation of gated neighbourhoods must surely inevitably lead to a decrease in solidarity between different social groups. The historical behaviour of the elites in Latin America today dominates also the middle- and lower middle-class populations. During the gating boom of the last two decades, neither Latin American media nor urbanists, architects or politicians have recognised it as a long-term problem for society. This is not surprising since all these groups, which are helping lead economic progress, directly or indirectly benefit from the gated communities. With the exception of Brazil and Colombia, gated neighbourhoods are partly a product of an imagined insecurity promoted by the media and real-estate companies, both gaining from exaggerating fears. On the other hand, as we have argued, residents may respond not so much to the security on offer as to the provision of neighbourhood services and facilities. Real-estate agents, developers, the media and an army of willing consumers have wielded sufficient economic power to prevent local and regional governments from attempting to regulate gated communities. Politics in Latin America are not driven so much by the implementation of citizens' desires as by the pressure of particular interests, and gated communities represent the perfection of Latin American traditions applied by the regional elite. Only recently and in the light of economic crisis is a rethink process starting, in which the negative consequences of urban disintegration through gated neighbourhoods have been raised. The announcement from the urban government in La Paz, Bolivia to open by force a dozen gated neighbourhoods to the public in March of 2003 shows that changes in urban policies may be possible in the future. Credible politicians and active civic participation are required to increase public awareness of urban social problems. And the problems being stored up by a fragmented urban form have not yet had the political debate in Latin America that they deserve.

Bibliography

Bähr, J. (1976) 'Neuere Entwicklungstendenzen lateinamerikanischer Großstädte', *Geographische Rundschau*, 28: 125–33.

Bähr, J. and Mertins, G. (1981) 'Idealschema der sozialräumlichen Differenzierung lateinamerikanischer Großstädte', *Geographische Zeitschrift*, 69 (1): 1–33.

Beckett, K. (1997) *Making Crime Pay: Law and Order in Contemporary American Politics*, New York: Oxford University Press.

Borsdorf, A. (1976) *Valdivia und Osorno. Strukturelle Disparitäten in chilenischen Mittelstädte*, Tübingen: Universität Tübingen (Tübinger Geographische Studien 69).

—— (1982) 'Die lateinamerikanische Großstadt. Zwischenbericht zur Diskussion um ein Modell', *Geographische Rundschau*, 34 (11): 498–501.

—— (1999) *Geographisch denken und wissenschaftlich arbeiten*, Gotha and Stuttgart: Klett.

—— (2002a) 'Barrios cerrados in Santiago de Chile, Quito y Lima: tendencias de la segregación socio-espacial', in L.F. Cabrales (ed.) *Latinoamérica: países abiertos, ciudades cerradas*, Guadalajara and Paris: Pandora: 581–610.

—— (2002b) 'Vor verschlossenen Türen – Wie neu sind die Tore und Mauern in lateinamerikanischen Städten? Eine Einführung', *Geographica Helvetica*, 57 (4): 238–44.

Borsdorf, A. and Hidalgo, R. (2005) 'Barrios cerrados y fragmentación social en América Latina. Estudio de las transformaciones socio-espaciales en Santiago de Chile 1990–2000', in R. Hidalgo, R. Trumper and A. Borsdorf (eds) *Transformaciones urbanas y procesos territoriales. Lecturas del nuevo dibujo de la ciudad latinoamericana*. Santiago de Chile: Geo-Libros: 105–22.

Borsdorf, A., Bähr, J. and Janoschka, M. (2002) 'Die Dynamik stadtstrukturellen

Wandels in Lateinamerika im Modell der lateinamerikanischen Stadt', *Geographica Helvetica*, 57 (4): 300–10.

Burgess, E.W. (1925) 'The growth of the city: An introduction to a research project', in R. Park, E.W. Burgess and R. McKenzie (eds) *The City*, Chicago, IL: University of Chicago Press.

Caldeira, T. (2000) *City of Walls: Crime, Segregation and Citizenship in São Paulo*, Berkeley, CA, Los Angeles, CA, London: University of California Press.

Ciccolella, P. (1999) 'Globalización y dualización en la Región Metropolitana de Buenos Aires. Grandes inversiones y reestructuración socioterritorial en los años noventa', *eure (Estudios urbano regionales)*, 21 (76): 5–28.

Coy, M. and Pöhler, M. (2002) 'Condomínios fechados und die Fragmentierung der Stadt. Typen – Akteure – Folgewirkungen', *Geographica Helvetica*, 57 (4): 264–77.

Crowley, W.K. (1995) 'Modelling the Latin American city', *Geographical Review*, 88 (1): 127–30.

Dammert, L. (2001) 'Construyendo ciudades inseguras: temor y violencia en Argentina', *eure (Estudios urbano regionales)*, 27 (82): 5–20.

Dear, M. (2000) *The Postmodern Urban Condition*, Oxford and Malden, MA: Blackwell.

Deler, J.P. (1992) 'Ciudades andinos: viejos y nuevos modelos', in E. Kringman Garcés (ed.) *Ciudades de los Andes: visión histórica y contemporánea*, Quito: Institut français d'études andines: 351–74.

ECIPSA (2003) *Valle Escondido News – La ciudad nueva*, Cordoba (publicity leaflet of Valle Escondido developers).

Ford, L.R. (1996) 'A new and improved model of Latin American city structure', *Geographical Review*, 86 (3): 437–40.

Gilbert, A. (1998) *The Latin American City*, 2nd edn, London: Latin American Bureau.

Gormsen, E. (1981) 'Die Städte in spanisch-Amerika. Ein zeit-räumliches Entwicklungsmodell der letzten hundert Jahre', *Erdkunde*, 35 (4): 290–303.

Griffin, E. and Ford, L. (1980) 'A model of Latin American city structure', *Geographical Review*, 70 (4): 397–422.

Harris, C.D. and Ullman, E.L. (1945) 'The nature of cities', *Annals of the American Academy for Political Science*, 242: 7–17.

Hidalgo, R. (1999) 'Continuidad y cambio en un siglo de vivienda social en Chile (1892–1998). Reflexiones a partir del caso de la ciudad de Santiago', *Revista Geográfica Norte Grande*, 26: 69–77.

Hoyt, H. (1939) *The Structure and Growth of Residential Neighborhoods in American Cities*, Washington, DC: Federal Housing Administration.

Janoschka, M. (2002a) *Wohlstand hinter Mauern. Private Urbanisierungen in Buenos Aires*, Vienna: Verlag der Österreichischen Akademie der Wissenschaften (Forschungsberichte des Instituts für Stadt- und Regionalforschung 28).

—— (2002b) 'El nuevo modelo de la ciudad latinoamericana: fragmentación y privatización', *eure (Estudios urbano regionales)*, 28 (85): 11–30.

—— (2002c) '"Stadt der Inseln". Buenos Aires: Abschottung und Fragmentierung als Zeichen eines neuen Stadtmodells', *RaumPlanung*, 101: 65–70.

—— (2002d) 'Urbanizaciones privadas en Buenos Aires: ¿Hacia un nuevo modelo de ciudad latinoamericana?', in L.F. Cabrales (ed.) *Latinoamérica: países abiertos, ciudades cerradas*, Guadalajara and Paris: Pandora: 287–318.

—— (2002e) 'Die Flucht vor Gewalt? Stereotype und Motivationen beim Andrang auf *barrios privados* in Buenos Aires', *Geographica Helvetica*, 57 (4): 290–9.

—— (2004) 'El modelo de ciudad latinoamericana. Privatización y fragmentación del espacio urbano de Buenos Aires: el caso Nordelta', in M. Welch (ed.) *La ciudad en cuestión. Nuevos lugares, viejos espacios*, Buenos Aires: Biblos (forthcoming).

Janoschka, M. and Glasze, G. (2003) 'Urbanizaciones cerradas: un modelo analítico', *Ciudades*, 59: 9–20.

Marcuse, P. (1989) '"Dual city": A muddy metaphor for a quartered city', *International Journal of Urban and Regional Research*, 13 (4): 697–708.

Meyer, K. and Bähr, J. (2001) 'Condominios in Greater Santiago de Chile and their impact on the urban structure', *Die Erde*, 132 (3): 293–321.

Nordelta (2002) *Nordelta CiudadPueblo – El camino para vivir mejor, Información*, Buenos Aires (publicity leaflet of Nordelta S.A., developer of Nordelta).

Ocampo, J. (2003) 'Las políticas económicas en América Latina y el Caribe', in Real Instituto Elcano (ed.) *Anuario Elcano: América Latina 2002–03*, Madrid: Real Instituto Elcano de Estudios Internacionales y Estratégicos: 260–75.

Parnreiter, C. (2004) 'Entwicklungstendenzen lateinamerikanischer Metropolen im Zeitalter der Globalisierung', *Mitteilungen der Österreichischen Geographischen Gesellschaft*, 146: 1–28.

Pöhler, M. (1999) *Zwischen Luxus-Ghettos und Favelas. Stadterweiterungsprozesse und sozialräumliche Segregation in Rio de Janeiro: Das Fallbeispiel Barra da Tijuca*, Tübingen: Universitätsverlag Tübingen (Kleinere Arbeiten aus dem Geographischen Institut der Universität Tübingen 21).

Reuband, K.-H. (1999) 'Kriminalitätsfurcht – Stabilität und Wandel', *Neue Kriminalpolitik*, 2: 15–20.

Roca, J. (2001) *Los Barrios Cerrados en Córdoba*, Córdoba: Universidad Nacional de Córdoba.

Rovira, A. (2002) 'Los barrios cerrados de Santiago de Chile: En busca de la seguridad y la privacidad perdida', in L.F. Cabrales (ed.) *Latinoamérica: países abiertos, ciudades cerradas*, Guadalajara and Paris: Pandora: 351–69.

Ruiz, J. (2003) 'Los siete pecados capitales de Iberoamérica: mito, realidad y consecuencias', in Real Instituto Elcano (ed.) *Anuario Elcano: América Latina 2002–03*, Madrid: Real Instituto Elcano de Estudios Internacionales y Estratégicos: 276–307.

Sabatini, F., Caceres, G. and Cerda, J. (2001) 'Segregación residencial en las principales ciudades chilenas: Tendencias de las tres últimas décadas y posibles cursos de acción', *eure (Estudios urbano regionales)*, 27 (82): 21–42.

Soja, E. (2000) *Postmetropolis: Critical Studies of Cities and Regions*, Oxford and Malden, MA: Blackwell.

Welch, M. (2002) 'Gartentürme des Wohlstands. Buenos Aires: Projektionen einer Wohnhaustypologie', *RaumPlanung*, 101: 71–6.

8

GATED COMMUNITIES IN SOUTH AFRICA

Ulrich Jürgens and Karina Landman

Abstract

In the course of the liberalisation and globalisation of South Africa, the transformation of the apartheid city to the post-apartheid city has contributed to an increase of insecurity. Recent statistics have shown comparatively high crime rates for inner city areas, both in terms of crimes against property and violent crimes. Unsurprisingly, the 1980s and especially the 1990s saw the proliferation of living areas in the suburbs, in particular, whose uniqueness and exclusiveness are defined by the degree of safety measures. In this chapter we discuss reasons for the popularity of so-called walled-in communities and explore the physical, social, political and urban planning consequences that follow from their development. Drawing on empirical research we identify those segments of the population that are most likely to buy into gated communities, ask if there are motives besides security, and consider the consequences of this type of settlement for the post-apartheid city.

Introduction

South Africa's apartheid cities, with their ideologically predefined segregation of residential areas according to ethnic criteria, underwent rapid change in the 1990s. The pluralisation of society had several profound implications for urban areas.

First, an almost unlimited spatial and social mobility has resulted in increasingly close spatial blends of ethnic groups. Second, not only has there been internal change in South Africa, but the country has become increasingly open to external influences. Now that boycotts and sanctions against the apartheid state have been lifted, the country is embedded in global economic networks and has turned into a magnet for migrants from many areas of Africa. Third, there has been a challenge to habits, social values and norms and to established patterns of behaviour and social status, i.e. new lifestyles occurred and the strong influence of conservative churches and ethnic barriers diminished. Therefore new forms of spatial and social fragmentation are emerging inside the post-apartheid city as the so-called gated communities.

This chapter sets out to explore the idea that gated communities are, in South Africa, a product of developments in post-apartheid society. It begins with an introduction to social changes that accompanied the transition in the late 1990s and early 2000s. The empirical research on which the chapter is based is briefly outlined and then we introduce the reader to the different types of gated community in South Africa. This is followed by a discussion of motivations for building and living in gated communities. Finally we consider the likely consequences of gating for the future of South African cities.

The ideas developed in this chapter seek to build on and probe the issues and propositions of other writers on South Africa's gated developments. Important among these is the idea that gated communities are a direct response to social pathologies (crime, perceived danger, poverty), but that they may be ineffective at countering or protecting from those pathologies and, worse, may produce their own negative consequences (Bremner 1999; Booysen 2001). The typology of South Africa's gated communities has only recently been addressed. The first comprehensive national survey was undertaken in 2002 (Landman 2003a). One aspect that almost all writers have addressed, whether the studies have been theoretical or empirical, is the likely impact and implications of gated communities in post-apartheid South Africa. These include crime reduction, crime displacement, spatial fragmentation, social exclusion and threats to local planning and policy and to democracy (Bremner 1999; Lipman and Harris 1999; Spinks 2001; Booysen 2001; Vrodljak 2001; Hook and Vrodljak 2002; Naude 2003; Jürgens, Gnad and Bähr 2003; Landman 2003a).

South African cities: from apartheid to post-apartheid

Physical changes

The main impact of the 1990s in respect of narrowing inequalities has probably been the improvement of service infrastructure in poor parts of South African cities, most notably in former black and coloured townships. Otherwise, the physical apartheid legacy inherited by the post-apartheid city has remained largely intact, i.e. former black areas still lie on the physical periphery of the cities. The major structural reforms required to alter the trajectory of urban change initiated in the 1980s did not take place. As a result, socio-spatial justice remains a major challenge in South Africa's increasingly fragmented cities, especially for those still subscribing to the egalitarian ethics (liberal or social) that underpinned much of the opposition to apartheid (Smith 2003: 30–1).

Spatially, apartheid cities were characterised by spatial fragmentation, segregation and low-density sprawl (Dewar 1992). Despite many efforts to address this, these patterns are still visible today. Gated communities have tended to reinforce these older patterns with new patterns of segregation (Bremner 1999;

Lipman and Harrris 1999; Vrodljak 2002; Jürgens and Gnad 2002; Harrison 2003). Tomlinson (2003: 86), commenting on the sharp distinctions and inequalities between the wealthy and poor areas in Johannesburg, warns of the divide between 'the walled residential communities and secure office parks and malls in the north [which] will stand in sharp contrast to the desperation of the south'. The socio-spatial segregation being reinforced by gated developments does not always, however, reflect the same spatial and social patterns as the segregation under apartheid. While apartheid segregation was primarily race-based, the socio-spatial segregation in the post-apartheid city is primarily income-based. The location of different types of gated community also depends on the type, the resources available and the specific physical characteristics of the environment.

Socio-economic changes

The sharp income and social differences between the population groups remained little changed after the abolition of apartheid. In 1996, 67.8 per cent of blacks, 14.3 per cent of Asians/Indians, 35.3 per cent of coloureds and 8.4 per cent of whites were classified as being poor. Expressed in standard living measures (SLM), which, as well as income, also measures standard-of-living indicators such as car, microwave and individual telephone ownership, the share of black persons in the lowest of eight LSM segments amounted to 99 per cent in 1998. In the highest segment, only 6.5 per cent were blacks (with 79.1 per cent whites) (SAIRR 2001: 372f.). Not only does the inequality between population groups remain high but the socio-economic gap within each group is also increasing.

The picture is not all bleak, however. There are also positive new developments. New opportunities for advancement in professional life have opened up, in particular for black persons. Between 1975 and 1996, their share of the group with the highest income increased by some 1,000 per cent (albeit starting from a very low level). By contrast, the white population's share of the highest income group decreased by 30 per cent while doubling in the lowest group to 5 per cent (SAIRR 2001: 376). While unemployment was almost unknown among this group under apartheid, the official numbers have increased to 7.8 per cent in 2002. This, however, still amounts only to one-third of the quota among coloureds and Asians/Indians, and only one-fifth of the unemployment share among blacks (SAIRR 2003: 161). Competition for well-paid jobs takes place against the background of a shrinking formal labour market and an increasing number of working people. The official unemployment rate for people under 25 varies around 40 per cent – this rate can be twice as high in peripheral areas. In the cities these developments have resulted in an increasing informalisation of the urban economy and an intensified, sometimes violent, competition for scarce housing and income resources.

Policy changes

The political transition in 1994 paved the way for the development of a range of new planning and development policies to address the imbalances of the past and pave the way for integration and socio-economic improvement in South African cities. Integration and development became key concepts in the new policies, setting out a new role for different spheres of government (Watson 2003). National government led the way by introducing a new law in 1995, the Development Facilitation Act (DFA), followed by the development of a Green Paper for Planning and Development (1999). The process finally culminated in the White Paper for Spatial Planning and Land Use Management in 2001. The white paper is based on a fundamental long-term vision and a set of principles and norms that calls for 'integrated planning for sustainable management of land resources' (White Paper on Spatial Planning and Land Use Management 2001: 2).

The new system of planning and land-use management consists of five essential elements: principles (achieving sustainability, equity, fairness and good governance), land-use regulations, IDP (integrated development planning); local spatial planning (a uniform set of procedures for land development approvals) and a national spatial framework. In addition, the attention of government and community planning institutions is no longer weighted towards well-to-do areas. Spending, investment, and police work tend to be equally directed towards the disadvantaged areas. This redirection of funds represents a redistribution of communal tax revenue. This, in turn, has led to increasing discontent with local politics and a retreat into privacy, particularly among the white population.

Perceptual changes

At the beginning of the 1980s, one of the important features of the South African city, aside from the apartheid zoning, was an unusually puritanical way of life. Exceptions to this were the inner-city areas of Johannesburg, especially Hillbrow and Yeoville, whose white inhabitants symbolised a highly liberal mentality. As a result, the influx of non-whites or mixed-race couples, though illegal until the repeal of the Group Areas Act, was initially considered an enrichment of the local culture and was explicitly welcomed. As in other *grey areas*, the pioneers consisted of well-educated coloureds, Indians and blacks on their way up the social ladder. Nevertheless, the proportion of non-white and mixed-race households, for example in Yeoville, was only 3.4 per cent in 1989. After the elections in 1994 the influx of non-white households increased, causing the proportion of this population group to rise to 84.1 per cent by 1998, and resulting in the perception of these areas by outsiders as ghetto-like and triggering fear that other areas may develop in the same way (Jürgens, Gnad and Bähr 2002, 2003). The socio-economic contrast among the inhabitants intensified and with it the area's social status. According to Saff (1995:

795), whites and middle-class blacks increasingly avoided the area: 'many whites are not opposed to living in the same neighbourhood as blacks as long as they perceive them to come from a similar class'.

Among the population of Yeoville, a climate of fear spread. Violent crimes, such as murder, armed robbery and rape, increased. The number of persons directly affected by crime is now very high by national and international standards. In every second white household in Yeoville at least one person reports being the victim of an attack. Among the non-white households, the figure is one in three. Physically, the areas are decaying due to overcrowding of housing units and slumlording (Gnad 2002). The social and physical decline has induced finance companies to deny mortgage loans by redlining – thus exacerbating the process of under-investment and physical degradation. As a result, property is sold at a great loss, if at all, and many owners have divided their houses or flats into several units, hoping to amortise the property in a short period of time through usurious rents. Associated with these processes are negative social changes – for example a noticeable increase in street prostitution, illegal taxi stands, car repair shops in backyards and so-called *shebeens* (houses or flats in which alcohol is sold).

The economic structure is adapting to the social transformation. Many restaurants and shops have closed. Since the mid-1990s, blacks from South Africa's neighbouring countries and from western and central Africa have been moving into this area and other parts of the inner city of Johannesburg (Morris 1998). Many are illegal migrants who are in danger of being deported. The street scene and especially the informal sector are increasingly being dominated by foreigners. Consequently, inner-city areas such as Yeoville are perceived as no-go-areas by the vast majority of whites and well-to-do blacks. In only one decade, the original residential population has retreated from much of the inner city to new living environments in the urban periphery. The perception of the inner cities as places of increasing insecurity is also shared by those who never lived there – an image propagated by the press and other media such as novels (Mpe 2001; Holland and Roberts 2002).

Empirical studies

In 1999 two gated communities were examined by the authors (Jürgens and Gnad 2002) in the Johannesburg metropolitan area. In terms of the typology developed by Blakely and Snyder (1997) both areas (Forestdale and Santa Cruz) are exclusively *security communities*. Based on information from the real-estate industry and the press, they show no signs of being *lifestyle* or *prestige* communities. Using a standardised questionnaire, households in 55 out of 60 completed and inhabited units in Forestdale, and 56 out of a total of 59 in Santa Cruz, were interviewed. Information about the ethnic composition of the remaining households was collected from neighbours. This was important since one of our principle research objectives was to investigate whether gated communities are a white phenomenon that revives apartheid-style residential segregation, or whether the affluent black population has also taken part in this new form of residential segregation.

A second study was conducted in 2002 by CSIR Building and Construction Technology (Boutek). This was an extensive study of gated communities in South Africa comprising two phases. The first focused on a national survey (quantitative approach) of gated communities to determine their extent and location. The second comprised four detailed case studies (qualitative approach) to assess the specific reasons for the spatial characteristics, the development and operation and the impact and implications of two types of gated community (enclosed neighbourhoods and large security estates) in two municipalities (cities of Johannesburg and Tshwane). The survey was conducted through questionnaires mailed to all the local and metropolitan municipalities (237) in South Africa. The case study research was conducted through semi-structured interviews, spatial analysis of the four neighbourhoods, direct observation in the neighbourhoods and document review. Interviewees included members of the boards of directors and residents' associations, as well as a small sample of 10 households who were willing to talk about their experiences of the areas. Interviews were also conducted with other players and stakeholders, such as the South African Police Service, the Metropolitan Police, local authority officials responsible for spatial planning and management, and the private security company operating in the area. The interviews were conducted towards the end of 2002 and the beginning of 2003. The document review covered a wide variety of sources, ranging from policies, council reports and technical notes about enclosed neighbourhoods, to documentation on the establishment and management of the enclosed areas obtained from residents and other involved parties. The latter included minutes of meetings, community newsletters, letters from residents to the committee and letters to council.

Gated communities in South Africa

Types, definitions and characteristics

Gated communities in South Africa can broadly be divided into two categories: security villages and enclosed neighbourhoods.

Security estates and town-house complexes

Security villages refer to private developments where the entire area is developed by a private developer. These areas or buildings are physically walled or fenced off and usually have a security gate or controlled access point, with or without a security guard. The roads within these developments are private and, in most of the cases, the management and maintenance is carried out by a private management body. Security villages include not only residential areas (such as town-house complexes and high-rise apartment blocks), but also controlled-access villages for business purposes (office blocks) and mixed-use

Figure 8.1 Cluster housing in Forestdale, Johannesburg. Photo: Ulrich Jürgens

developments, such as large security estates. Discussion in the remainder of this chapter addresses two types in particular, namely town-house complexes and large security estates.

Town-house complexes are residential areas dominated by terraced houses. In contrast to classical single houses with surrounding free space, residents gather together closely, in a literal sense, the better to control the surrounding areas together. Residents sacrifice anonymity and individual garden space for multi-storey buildings and communal spaces and facilities. This construction type, also known as cluster housing, appeals to middle-class customers. Instead of image and prestige, a pragmatic desire for more secure living is the underlying motivation. Town-house complexes can be of very different sizes; they are also more suitable than detached housing developments for redeveloped land in older residential areas.

Large security estates in South Africa are mostly located on the urban periphery where bigger portions of land are available, as well as pre-existing design elements such as rivers, dams and trees. They offer an entire lifestyle package, including a secure environment; a range of services (garden services, refuse removal, etc.); and a variety of facilities and amenities such as golf courses, squash courts, cycle routes, hiking routes, equestrian routes and water activities. Most of the luxury security estates occupy between 10 and 50 hectares, while two of the more ambitious estates occupy larger areas, namely Heritage Park in the City of Cape Town (200 hectares) and Dainfern in the City of Johannesburg (350 hectares).

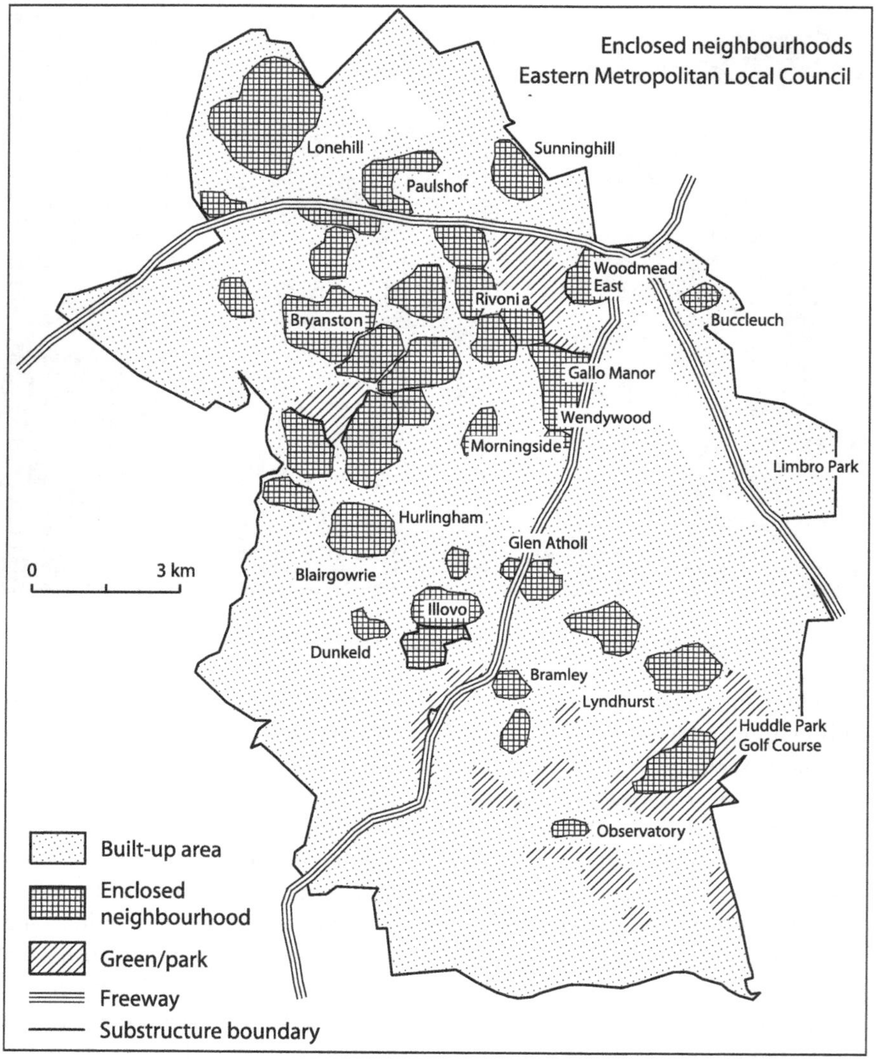

Figure 8.2 Gated communities in the eastern part of Johannesburg. Source: Landman 2000

Enclosed neighbourhoods

Enclosed neighbourhoods are existing neighbourhoods that have controlled access through gates or booms across existing roads, showing a mosaic-like pattern in areas like eastern Johannesburg (figure 8.2). Many have also been fenced or walled off, with a limited number of controlled entrances and exits and private security guards stationed at these points or even inside the scheme. The roads within these neighbourhoods were previously, or still are, public property, depending on the model used within different local authorities. In

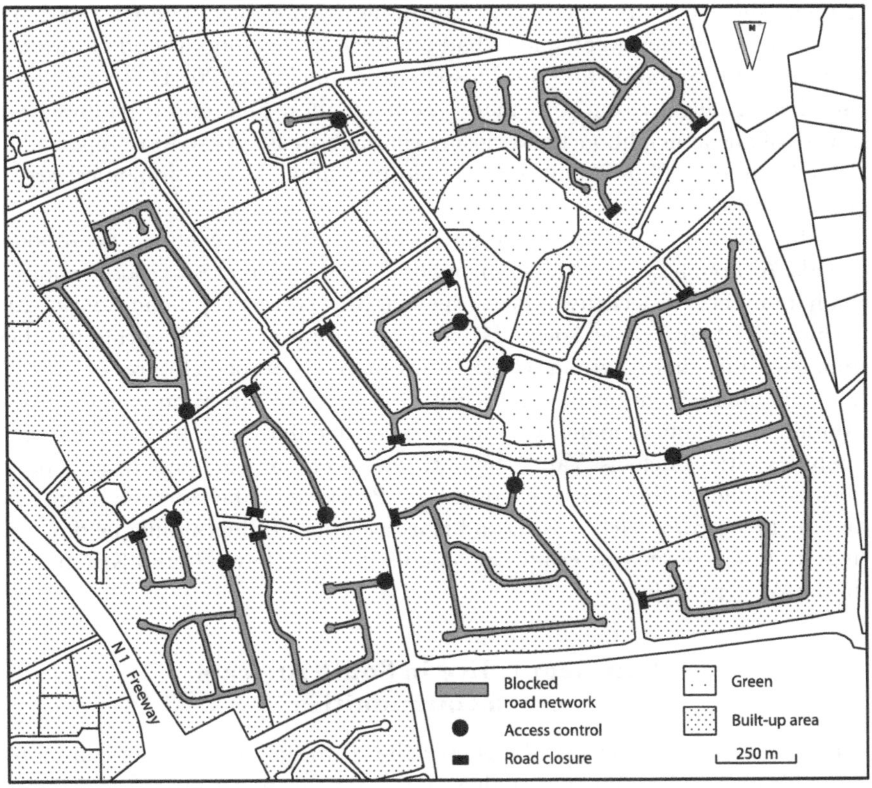

Figure 8.3 Map of enclosed neighbourhood, Fourways, Johannesburg. Source: survey University of Kiel 1999

the majority of enclosed neighbourhoods roads remain publicly owned behind the barricades (figure 8.3).

Diffusion of gated communities in South Africa

The national survey confirmed the widespread presence of both types of gated community in South Africa and showed that the most common type (35 per cent), in the municipalities that responded, were town-house complexes.

The numbers of enclosed neighbourhoods and security villages vary substantially from one municipality to the other. This is true for both the numbers and proportions of each type. The survey[1] indicated the highest numbers of enclosed neighbourhoods were in Gauteng, with two municipalities having from seven to nine neighbourhood closures, two having from 16 to 25, one from 25 to 100 and one had more than a hundred. Two of the three metropolitan municipalities had the highest numbers of enclosed neighbourhoods at the time of the survey – namely Tshwane with 36, and Johannesburg with 300. There were also two municipalities in the Western Cape with higher numbers, namely those that included Cape Town and Mossel Bay. The City of Cape Town metropolitan municipality had recorded 25 neighbourhood closures at the time of responding and the Mossel Bay municipality 20. There are also a number of illegal closures in many municipalities. Five municipalities were aware of illegal neighbourhood closures in their areas. There were reported to have been more than 200 illegal road closures in the City of Johannesburg in January 2003.

The distribution of security estates differs to that of enclosed neighbourhoods. Although some of the municipalities facing road closures also reported high numbers of security estates, such as the cities of Johannesburg (20), Tshwane (18) and Cape Town (24), high numbers were also reported in other municipalities that recorded relatively low numbers of road closures or none at all. The highest numbers of security estates were in Emfuleni (40) and Madibeng (31) municipalities. Emfuleni is located in the south of Gauteng and includes cities such as Vanderbijlpark and Vereeniging, and Madibeng is located in the south-east of the North West Province on the Gauteng boundary, and includes towns such as Brits and Hartbeespoort. Other municipalities with high numbers of security estates include Plettenburg Bay (21) and Knysna (20) in the Western Cape.

Why people live in South Africa's gated communities

There are a host of reasons why people choose to stay in the gated communities of South Africa. These include safety and security, a sense of community and identity, the appeal of shared social values, homogeneity and control, economic control, specific lifestyle, fashion, the efficiency and effectiveness of neighbourhood management and governance, and status and prestige. We focus on three of these in more detail.

Safety and security

CSIR's research, based on four case studies, (Landman 2003b–e) has indicated that, while not the only reasons, safety and security and fear of crime are the main reasons behind the proliferation of road closures and security estates. Only a few incidents in a neighbourhood are enough to aggravate fears of crime and victimisation. People start to talk about crime every day and speculate about ways to reduce the risk. As a result, community action is initiated and in some instances this leads eventually to road closure. Without doubt, therefore, the fear of crime plays a significant role in the drive for neighbourhood closure. One particular aspect of this is the fear of being hijacked (despite extensive target-hardening measures).[2] As one resident of an enclosed neighbourhood in Johannesburg explained,

> When we moved in here we just had chicken-wire fencing around most of the property, and just steel tubing and a wire gate. Then eventually we upgraded that to a proper wall and a decent gate. It was still not a locked gate. It was only some time after that we went the route which you find now . . . did you see – most people have electronic gates? So eventually we went to that [electronic gates] because . . . my wife's cousin from Cape Town . . . they actually stole his combi from our driveway. This was before the street closure. So then we thought we must at least do something so then [we] put in an electric gate. But even then it got to the point where . . . especially at night, when you reversed out of your driveway and got out to open the gate, there was a certain level of nervousness about reversing into the street [. . . and getting] mugged . . . and after dark [if I saw a] car's headlights behind [me] and it would follow me into the area into our street, more than once I would take a drive around the blocks to see how far this car was going to follow me. But . . . definitely . . . you had that level of nervousness.

People were nervous and applied a number of preventive measures, from physical target-hardening to social control – driving around the block to check out a suspect. It was not a case of immediately going for the extreme option of closing off the entire neighbourhood; the latter came only after a gradual build-up. Residents systematically upgraded the security measures around their houses, until they came to a point where they had to secure the neighbourhood as well. It was an evolutionary process of gradually securing their surroundings, from small alterations to the individual house, to bigger alterations to the property, to finally transforming the entire neighbourhood.

Safety and security was also a major driver behind the choice to reside in a security estate. A statement from a resident illustrates the point:

> Well, as you say, first and foremost for the security and in our case the golf course . . . but from a security point of view I think that was the main driving force of the thing and the golf course a secondary reason . . .

All those interviewed in this particular security estate in the City of Tshwane maintained that safety and security were their major motives for living on the estate, followed by lifestyle choices.

This also appears to be the case in town-house complexes. Residents of these seem also aware of the price they have to pay for security. Some 60 per cent of households in Forestdale and about 45 per cent in Santa Cruz report that they have accepted the fact that, compared to their former homes, the average number of rooms has decreased. Most of them had previously lived in single houses in the Gauteng metropolitan area.

Other reported reasons for moving into gated communities are their central locations (proximity to schools, the motorway, shopping centres and to the workplace), affordability and the immediate social environment ('quiet area', 'upmarket area'). These are typical criteria influencing location choice, which operate in combination with the security factor.

Internal/external population

Neither within nor without the compounds is the acceptability of walled communities undisputed. The high building density of cluster settlements in particular is criticised because this is considered to be detrimental to the prestige of the surrounding communities and thus to reduce the value of real estate there. Comments such as the following are common: 'This is a downgrading of the area and a traffic hazard.' 'Dense population of our area will encourage crime and lower our property values.' A black homeowner from outside the gated community wrote:

> I moved from Soweto to get a peace of mind. Unfortunately, I find that there is another Soweto with matchbox houses to be built nearby. I strongly oppose this downgrading development to our upmarket sizes and larger stands size of extension 66.
>
> (Residents in Douglasdale, Johannesburg, 1994)

Reacting to road closures, a civil rights group (the Open City Forum) has been formed by citizens who feel restricted in their individual freedom of movement.

Within the gated communities the lack of privacy, due to light construction and resulting poor acoustic insulation, as well as the narrow space between the houses, is frequently criticised. Although crime in some communities seems to have been excluded, fears of the threat of violence outside the fences are not.

Organisation

An important aspect for the functionality of security villages is their internal organisation. In accordance with the Sectional Titles Act of 1986 the property owners and tenants of a compound are obliged to abide by conduct rules. In Santa Cruz, even the short version of conduct rules comprises 19 points, listing prohibited behaviour and illegal modifications to the buildings. Trustees and a chairman serve as representatives and spokesperson respectively of the community; in Forestdale there is a homeowners' association. They are obliged to balance their decisions between the private interests of individuals and the interest of the community as a whole. In Santa Cruz, for example, many internal services such as paying electricity bills to the city, the control of conduct rule observance, as well as installation and repair works in the compound, lie in the hands of the committee. These services are paid for with money from mandatory charges that are collected from all households in addition to the regular electricity bills. Homeowners' associations (HOAs) also play a very important role in the daily management of large security estates, enforcing adherence to a large number of strict rules, regulations and controls (RRCs) linked to ownership and residency inside these estates.

Committee-like structures also evolve in the case of enclosed neighbourhoods. Their main objective is to distribute costs evenly and fairly among the (voluntary) participating households. Here, peer pressure among the neighbours makes it quite difficult to evade majority initiatives. Another problem for committee structures and their future planning is the loss of paying members which always happens when they sell their houses or suffer short-term financial constraints (Ferndale Village Scheme 2003).

The impacts of gated communities

Physical impacts

While one specific enclosed neighbourhood may not have a material impact on the overall urban form (due to its layout, location in the road network and size), several can lead to significant urban transformation (figure 8.2). The big question is therefore a systemic one: is it possible to achieve city-wide planning goals when neighbourhoods are becoming increasingly impermeable? The majority of planning policies in South African cities, for example, seek to promote integrated public transport systems that link a network of corridors and nodes and provide easy access to sustainable neighbourhoods. Such visions seem likely to be made that much more difficult – possibly impossible in their current form – by the multiplication of enclosed neighbourhoods in, for example, Johannesburg and Tshwane. The sheer numbers of these types of development suggest that they are not likely to decrease in the immediate future. There is therefore a real prospect of the cities of the future being shaped by a number of isolated super-blocks linked by a network of rapid road-based transport corridors. Such spatial fragmentation carries real risks of social separation that need to be taken seriously.

Large security estates also have a number of spatial implications. Dainfern (northern Johannesburg), for example, comprises an area of 320 hectares. This area is accessible only to those residing within it, or with a reason to be there (visitors, contractors and so on). To all other urban residents the area is inaccessible. As such, it has created a huge impenetrable super-neighbourhood in the urban landscape. This is contrary to the idea of integrated development promoted by the City of Johannesburg. It contributes to the old pattern of inappropriate densities, i.e. low-density suburban sprawl and fragmented development (due to separate neighbourhoods for separate race groups, divided by buffer strips and rapid transport routes or railway tracks), albeit for the main purpose of crime prevention. This contradicts the current vision set out in the White Paper on Spatial Planning and Land Use Management 2001, advocating neighbourhood and metropolitan spatial integration. When Dainfern was initiated in 1991, it was the only one of its kind in Johannesburg. However, since then, similar developments have proliferated across the city and across other South African cities. While one can argue that one or two luxury estates may be tolerated to offer a choice of lifestyle to the rich, a large number can seriously compromise the aims set out in the local spatial development frameworks (SDFs), based on the National Spatial Development Framework. The precedent set by these developments may mean that the wholesale privatisation of higher-income suburbs is difficult to stop. This would tend to undermine adopted planning strategies.

Having said this, the needs and perceptions of the city's occupants will continue to drive responses in the physical environment. Johannesburg, as many other larger cities, is perceived to be an extremely unsafe place to live and the fear of crime is very high. A single incident in a neighbourhood can raise panic among residents. This raises the idea that the goal of an 'integrated city' – the basis of much urban policy in South Africa – may be unrealistic. Alternatively, perhaps what is needed is a more detailed exploration of how to achieve integration and make it work for everybody in the city – including those who demand greater security. For example, should walls and fences be endured temporarily to enhance safety and security, despite their impact on the urban form? Or should planners take a stronger stance, opposing the increased spatial fragmentation and separation, and work with specialists to find friendlier methods of crime prevention, including preventive planning and design?

Socio-economic impacts

All over South Africa, gated communities have influenced prices of flats and houses, which increasingly depend on the extent of security measures. Especially in prestige communities, where security and lifestyle are two sides of the same coin, land and house prices have rocketed, notwithstanding fears at the top end of the market that cluster-settlements may perversely increase crime by design ('in a cluster development you can be exposed to the comings and goings of everyone else's staff. And it's a fact that a high percentage of burglaries are inside jobs'; Stafford 1999: 46). Particularly for the white and black middle class, for first-time homeowners and single households fleeing urban blight and their inner city flats, gated communities are becoming the most important housing alternative 'in desirable areas'. It remains unclear whether the social segregation in these compounds is complemented or eclipsed by a new form of racial segregation, which comes about as a result of different price levels and images of individual residential areas. Changes that can already be observed in Forestdale and Santa Cruz do not exclude the possibility that the various ethnic communities in South Africa will eventually set up their 'own' gated communities (figure 8.4). 'They slaughter chickens in their gardens and disturb regardless of their neighbours with noisy parties. Empty cans are dumped in the environment and they used to pee outside in the complex . . .'; all of that is regarded as township culture and rejected by white neighbours (interviews conducted in 1999 by Gnad; in Gnad 2002).

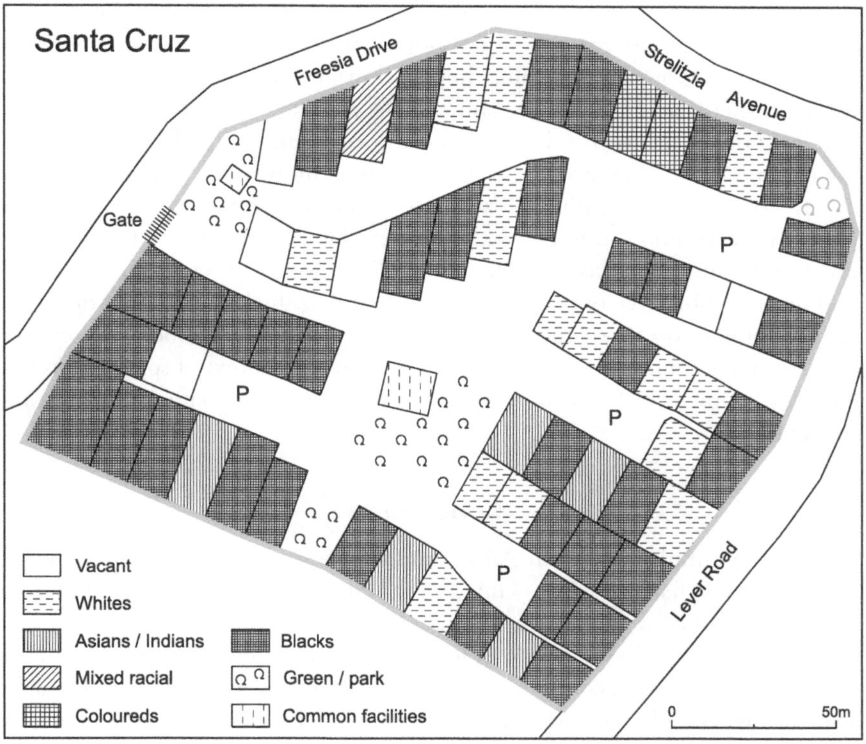

Figure 8.4 Ethnical structure of inhabitants in a cluster settlement. Source: survey University of Kiel, 1999

Institutional impacts

A number of questions become relevant when considering the implications of gated-community policy formulation and institutional management. The main question relates to where (literally) to draw the line. Should those areas that lend themselves to neighbourhood closure be allowed to close their roads, while others which are located unsuitably be refused permission to do so? Or should all applications be refused because this would be fair to those who are not in a position to close off their roads? And what if they then went ahead and closed off their roads illegally at huge expense? Furthermore, how should the council deal with angry residents complaining about the high crime rate in their areas, or with residents who maintain that their area is suitable for road closures?

One cannot therefore consider a single development in isolation. One should also consider its implications for policy formulation and wider urban development. Should areas with those who can afford to pay for certain services and additional resources be able to do so, regardless of the impact on the future sustainability of the city? This may suggest that the development of different types of gated community should be considered when developing future SDFs and urban management plans. Perhaps there should be a gated-community policy to govern the development of such estates. Large security estates can have significant policy implications, even more so if their size and numbers increase in the future. Some of the aspects that should be dealt with in a local policy are access, impact on traffic and suitability of neighbourhood for enclosure. Policies would need to consider more than just the technical closure of a road or gating of an area formerly permeable to the general public; they would need to consider the overall impact on the spatial form and functioning of the city as a whole. The policy should be set within the larger and primary objective of reducing crime. Is gating the best and only option to reduce crime? Are there alternatives? Enclosed neighbourhoods and security developments cannot be considered only within the context of spatial planning and development policies; they are part of the larger debate on local crime prevention.

Conclusion

Gated communities in South Africa are clearly a response to high levels of crime and the fear of crime. Gated communities are not only restricted to white higher-income areas, but also occur in middle-income areas and have mixed-race groups living inside. They are indeed a new form of segregation but it is a social and economic segregation rather than a new form of racial segregation. Significant numbers of the new black elite have also taken residence inside these communities, responding to the fear of crime and of 'others' by seeking to exclude them. South Africa's gated communities may be thought of as creating and preserving new socio-spatial groupings, which may be socially homogeneous, culturally homogeneous or income-homogeneous (but racially mixed). South Africa therefore faces the difficult challenge of eradicating the spatial divisions of apartheid while responding to new demands for income-based segregation. Balancing the tensions between the short-term need for safety, security and efficient service delivery with the longer-term goals of integrated development and equity is not easy. A starting point for constructive discussion would be to disentangle the debates about old-style (racial) spatial segregation and new-style (income) spatial segregation. Given the depth of feeling among South Africans about the former issue, this may be a tall order, however.

Notes

1 As mentioned earlier, the survey was conducted in 2002. The numbers of gated communities have increased in some areas since then, notably Johannesburg and Tshwane.

2 Target-hardening is the physical strengthening of building façades or property boundaries, for example through burglar bars in front of windows, security gates in front of doors and boundary walls or fences.

Bibliography

Blakely, E. and Snyder, M. (1997) *Fortress America: Gated Communities in the United States*, Washington, DC: Brookings Institution Press.

Booysen, A. (2001) 'Enclosed neighbourhoods in perspective', unpublished hounours thesis, University of Pretoria.

Bremner, L. (1999) 'Crime and the emerging landscape of post-apartheid Johannesburg', in H. Judin and I. Vladislavic (eds) *Blanc Architecture: Apartheid and After*, Rotterdam: Nai Publishers.

Dewar, D. (1992) 'Urbanization and the South African city: a manifesto for change', in D.M. Smith (ed.) *The Apartheid City and Beyond: Urbanization and Social Change in South Africa*, London: Routledge.

Donaldson, R., Jürgens, U. and Bähr, J. (2003) 'Inner-city change in Pretoria: social and spatial trends', *Acta Academica Suppl.*, 1: 1–33.

Ferndale Village Scheme (ed.) (2003) *The Village Voice*, November, Randburg.

Gnad, M. (2002) *Desegregation und neue Segregation in Johannesburg nach dem Ende der Apartheid*, Kiel: Kiel University Press.

Harrison, P. (2003) 'Fragmentation and globalisation as the new meta-narrative', in Philip Harrison, Marie Huchzermeyer and Mzwanele Mayekiso (eds) *Confronting Fragmentation: Housing and Urban Development in a Democratising Society*, Cape Town: University of Cape Town Press: 13–25.

Holland, H. and Roberts, A. (2002) *From Jo'burg to Jozi*, Johannesburg: Penguin Books.

Hook, D. and Vrodljak, M. (2002) 'Gated communities, heterotopia and a "rights" privilege: A "heterotopology" of the South African security-park', *Geoforum*, 33: 195–219.

Jürgens, U. (2002) 'South African inner city areas in comparative perspective – social and economic transformation in Johannesburg, Bloemfontein, Pretoria, and Durban', paper presented at the 18th International Conference of the Africanist Organisation in Germany, Hamburg, May 2002. Online. Available HTTP: http://www.vad-ev.de/papers/paper_juergens.pdf (accessed 29 July 2005).

Jürgens, U. and Gnad, M. (2002) 'Gated communities in South Africa – experiences from Johannesburg', *Environment and Planning B: Planning and Design*, 29: 337–53.

Jürgens, U., Gnad, M. and Bähr, J. (2002) 'Residential dynamics in Yeoville, Johannesburg in the 1990s after the end of apartheid', in A. Osmanovic (ed.) *Transforming South Africa*, Hamburg: Institut für Afrikakunde: 172–206.

—— (2003) 'New forms of class and racial segregation: Ghettos or ethnic enclaves?', in R. Tomlinson, R. Beauregard, L. Bremner and X. Mangcu (eds) *Emerging Johannesburg – Perspectives on the Postapartheid city*, London: Routledge: 56–70.

Jürgens, U., Marais, L., Barker, C. and Lombaard, M. (2003) 'Socio-demographic transformation in the Bloemfontein inner-city area', *Acta Academica Suppl.*, 1: 34–54.

Landman, K. (2000) *An overview of enclosed neighbourhoods in South Africa*, Pretoria: CSIR Publication.

—— (2003a) *Gated communities in South Africa: A National Survey*, Pretoria: CSIR Publication.

—— (2003b) *Case Study Research on Gated Communities in South Africa: The Case of Gallo Manor, Johannesburg*, Pretoria: CSIR Publication (reference number BOU/I 253).

—— (2003c) *Case Study Research on Gated Communities in South Africa: The Case of Dainfern, Johannesburg*, Pretoria: CSIR Publication (reference number BOU/I 254).

—— (2003d) *Case Study Research on Gated Communities in South Africa: The Case of Strubenkop, Pretoria*, Pretoria: CSIR Publication (reference number BOU/I 255).
—— (2003e) *Case Study Research on Gated Communities in South Africa: The Case of Woodhill, Pretoria*, Pretoria: CSIR Publication (reference number BOU/I 256).
Leggett, T. (2001) *Rainbow Vice – The Drugs and Sex Industries in the New South Africa*, Cape Town: David Philip Publications.
Lipman, A. and Harris, H. (1999) 'Fortress Johannesburg', *Environment and Planning B: Planning and Design*, 26: 727–40.
Morris, A. (1998) 'Our fellow Africans make our lives hell: The lives of Congolese and Nigerians living in Johannesburg', *Ethnic and Racial Studies*, 12: 1116–36.
Mpe, P. (2001) *Welcome to our Hillbrow*, Pietermaritzburg: University of Natal Press.
Naudé, B. (2003) 'The effectiveness of public road closures in suburban areas as a crime reduction measure', *Security Focus*, July: 34–6.
Saff, G. (1995) 'Residential segregation in postapartheid South Africa – what can be learned from the United States experience', *Urban Affairs Review*, 30: 782–808.
SAIRR (South African Institute of Race Relations) (2001) *South Africa Survey 2000/2001*, Johannesburg.
—— (2003) *South Africa Survey 2002/2003*, Johannesburg.
Shaw, M. (2002) *Crime and Policing in Post-apartheid South Africa: Transforming under Fire*, Cape Town: David Philip.
Smith, D. (2003) *Urban Fragmentation, Inequality and Social Justice: Ethical Perspectives*, Cape Town: University of Cape Town Press.
Spinks, C. (2001) 'A new apartheid? Urban spatiality, (fear of) crime and segregation in Cape Town, South Africa', *London School of Economics Working Paper Series*, 1–20: 1–42.
Stafford, L. (1999) 'The return of the big house', *Personal Wealth*, 2nd quarter: 45–8.
Statistics South Africa (2001) *South Africa in Transition*, Pretoria.
Thulare, P. (1999) 'Welcome to Hellbrow – talking to Johannesburg's drug syndicates', *Crime and Conflict* (Indicator SA), 16: 17–21.
Tomlinson, R. (2003) 'HIV/Aids and urban disintegration in Johannesburg', in P. Harrison, M. Huchzermeyer and M. Mayekiso (eds) *Confronting Fragmentation: Housing and Urban Development in a Democratizing Society*, Cape Town: University of Cape Town Press: 76–87.
Turok, I. (2001) 'Persistent polarisation post-apartheid? Progress towards urban integration in Cape Town', *Urban Studies*, 38: 2349–77.
Vrodljak, M. (2002) 'Place and the politics of subjectivity', unpublished master's thesis, Johannesburg: University of the Witwatersrand.
Watson, V. (2003) 'Planning for integration: The case of metropolitan Cape Town', in P. Harrison, M. Huchzermeyer and M. Mayekiso (eds) *Confronting Fragmentation: Housing and Urban Development in a Democratizing Society*, Cape Town: University of Cape Town Press: 140–53.

9

THE SPREAD OF PRIVATE GUARDED NEIGHBOURHOODS IN LEBANON AND THE SIGNIFICANCE OF A HISTORICALLY AND GEOGRAPHICALLY SPECIFIC GOVERNMENTALITY[1]

Georg Glasze

Abstract

Since the 1980s privately organised guarded apartment and villa complexes have mushroomed in the areas around Beirut and in the developing suburban fringe south of Tripoli in North Lebanon. The distinct nature of these developments raises the question of their causes. On the one hand, some Lebanese commentators see them as a direct consequence of the war or interpret them as 'traditional' ethnic and confessional segregation in a new guise. On the other hand, the international scientific discussion on the spread of this kind of housing in many regions of the world is dominated by universalistic explanations. The framework I adopt in analysing the Lebanese case is sympathetic to both approaches. Its focus is on urban governance and it considers the segmented patterns of social interaction that provide a sine qua non context for the development of guarded residential complexes in Lebanon during two specific periods.

Introduction

Lebanon's private guarded neighbourhoods have mushroomed since the 1980s in the surroundings of the capital Beirut and in the developing suburban fringe south of Tripoli in North Lebanon. They are of two types: *condominiums* containing apartments (see figure 9.1 overleaf) and *gated settlements* with dominantly single-family homes or terraced houses.

While most of these complexes are used as a main residence, there are also

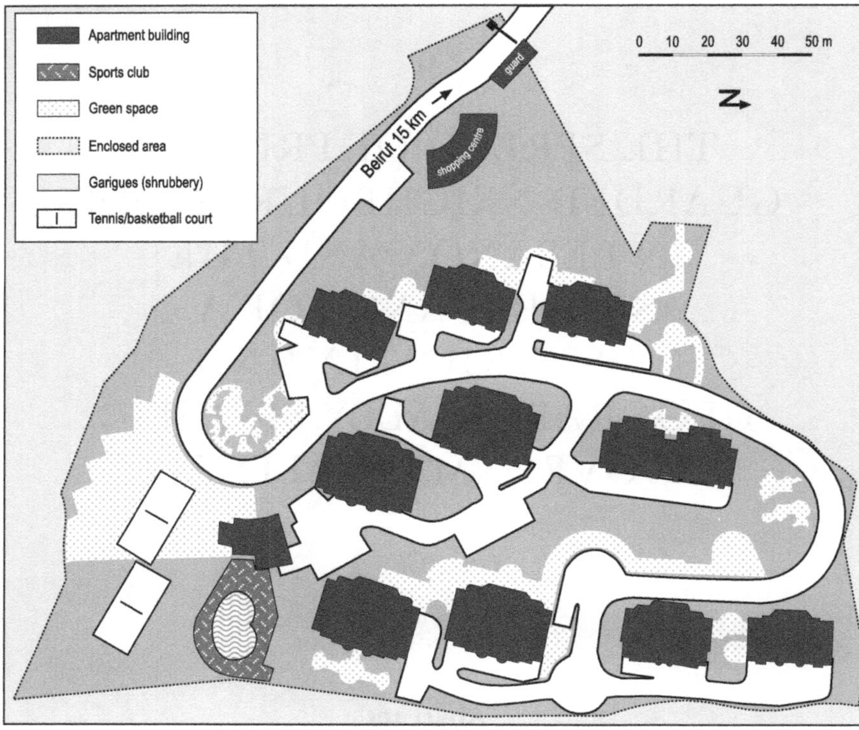

Figure 9.1 The condominium Al Majal established in the 1990s on a hill overlooking southern Beirut. Source: Glasze and Alkhayyal 2002

several gated beach and mountain *resorts* as well as two gated ski villages (see figure 9.2 opposite), mostly used as secondary residences.

All these projects combine common property with the individual property of a housing unit, specifically the right to use an apartment or house plus shared facilities, with some kind of access limitation through gates, fences or walls. They offer maintenance and 24-hour security services as well as natural (e.g. beach, view) and artificial (e.g. pool, tennis court) amenities. The common property and the common services are managed by a self-governing organisation. Altogether these estates represented approximately 2 per cent of the total Lebanese housing supply in 2000 and an estimated 7 per cent of the housing stock built during the 1980s and 1990s.

The spread of this kind of neighbourhood throughout the world has triggered a new discussion about the relationship between social and urban development. This has been focused chiefly on the US and has been marked by a certain US-centrism and by universal approaches seeking to abstract from the US experience. Observations from the US have been taken as indications of global trends, often without basing these assumptions on empirical work.

Some of the few Lebanese authors who have discussed the trend have linked the establishment of private guarded neighbourhoods in their country not to

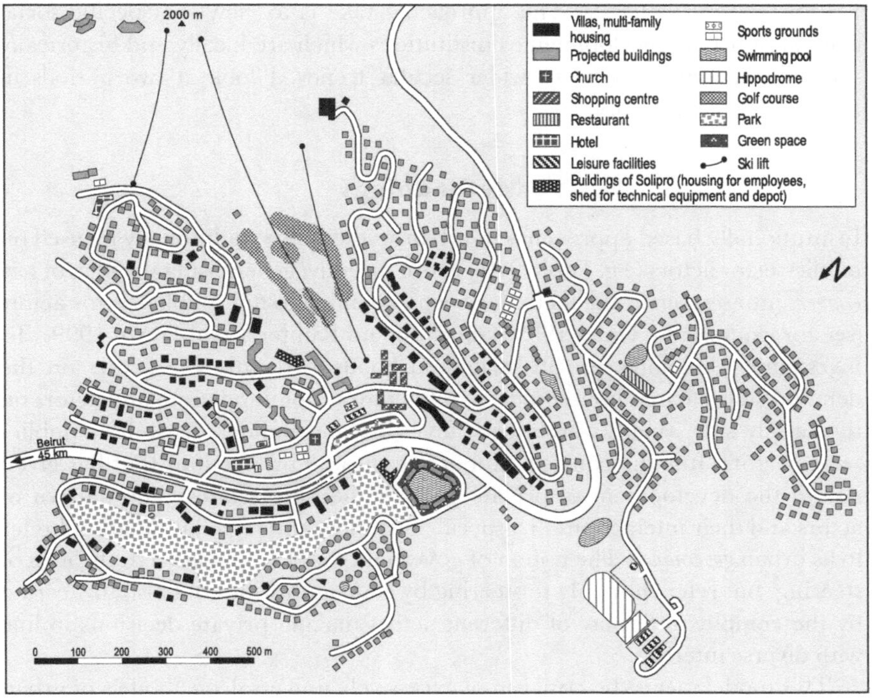

Figure 9.2 The gated ski village Faqra in the high mountains of Mount Lebanon – a playground of the Lebanese elite. Source: Glasze 2003

global but to specific local processes. Several writers interpreted these 'exclusive "bubbles" . . . of self-sufficiency' (Tuéni 1998: 285) or 'autonomous complexes' (Davie 1994: 61) as a result of the Lebanese war. The Lebanese political scientist Michael Young sees these 'new cantonments' as a new retreat to 'the mountains' and a rejection of the city as the place of living together (1995). He links the boom of private neighbourhoods in post-war Lebanon to the traditional patterns of ethnic and confessional segregation.[2] This is an important conjecture arising from the observation that spatial seclusion of social groups is not a new phenomenon in the cities of the Arab world. Urban research on pre-modern towns depicts the socio-spatial and material fragmentation of urban patterns into small and distinct quarters as one of the most typical characteristics of Arab cities (Abu-Lughod 1987; Raymond 1989; Wirth 2000). Raymond describes the 'compartmentalization of the city . . . [being] particularly marked in the case of numerous religious and ethnic communities' (1994: 15). However, it would be over-simplistic to describe the modern developments in Lebanon as just a reappearance of the fragmented settlement patterns in many of the old towns (Glasze and Alkhayyal 2002). On the other hand, the universalistic approaches, which dominate international scientific discussion, do not arguably facilitate a sufficient analysis of the specific contexts in which this kind of segregated housing pattern emerges. The approach I take is to view the specific social context of urban development as institutions which are locally and historically specific, but not immune to wider secular trends. I look at two periods in Lebanon's recent history.

Methodology

Institutionally based approaches in urban studies have traditionally focused on public-sector actors (e.g. Pahl 1970). More recently, generalising the idea of *government* into *governance*, the focus has been expanded to non-public-sector actors (see for example Le Galès 1995; Goodwin and Painter 1996; Pierre 1999). To investigate Lebanon's private neighbourhoods I identify the actors on the demand side, such as households searching for accommodation, and actors on the supply side, such as architects, investors and builders as well as public-sector actors. It is the constellation of actions and interactions that has given rise to the development of private guarded housing estates. The pattern of actors and their interactions in a specific historical and regional situation I refer to as urban *governance*. The notion of governance is closely related to the idea of steering, but refers not only to steering by state authorities but also to steering by the complex interplay of different actors making private decisions in line with diverse interests.

This model meets the criticism of excessively universalistic models of urban development; the notion that 'politics matter' is conceptualised by taking into account not only the actors of the demand and supply side but also the actors with political power as well as the 'rules' of their interactions. The dictum that 'history matters' is conceptualised as the 'third dimension' of institutions. Defined as supra-individual patterns of human interaction, *institutions* create what is called path-dependency. And, as the rules of the interplay of different actors are often solidified as institutions which refer to specific territories – an extreme example are formal institutions such as laws and decrees – this approach enables us to conceptualise regionally different contexts of urban development as a 'territorial arena', which means that 'geography matters' (for a detailed discussion of the conceptual basis and its relation to institutional theory and 'new institutional economics' see Glasze 2003: 43–9).

The case study is based on a triangulation of methods of qualitative and quantitative social studies as well as classical geographical research (Glasze 2003). In view of the absence of even elementary empirical information on the phenomenon, the starting point was to map Lebanon's private neighbourhoods. This was followed by narrative interviews with 25 inhabitants; three group interviews with inhabitants; participant observation in two neighbourhoods; 50 problem-oriented interviews with experts including architects, developers, representatives of the media, advertising agencies, private security services, environmental NGOs, employees of public planning bureaux and two mayors of municipalities affected by guarded neighbourhoods. I also conducted qualitative content analysis of media coverage, laws and decrees, planning concepts and advertising in brochures and on the Internet. These qualitative methods were complemented by a standardised survey of 653 inhabitants in a dozen complexes, which helped to retrieve information about demographic and socio-economic structures of the households within guarded housing estates. It also yielded important information about the inhabitants' housing biographies as well as about the community life within the complexes.

From enclaves of well-being to lifestyle enclaves

Private neighbourhoods in Lebanon have mostly been established in two periods. During the civil war, from 1975 to 1990, the first resorts and condominiums began to appear. In a second phase, during the 1990s, many condominiums as well as a few resorts and gated settlements sprang up in the surroundings of Beirut, Tripoli and Saida.

Exclusive refuge: resorts and condominiums during the civil war

The construction of the first resorts and condominiums in Lebanon is closely related to the specific context of the civil war and was concentrated to a high degree on the Kesrouane region in the centre of the area which was controlled by Christian militias – the so-called Christian canton. Here, a specific form of cooperation was established over several years between the resort and condominium developers, the leaders of the militias and families looking for new housing – mostly Christians taking refuge in the region (see figure 9.3).

In view of the missing foreign tourists and the influx of Christians from Beirut and Tripoli some hotel owners on the coast of the Kesrouane region decided in the mid-1970s to transform their building complexes from hotels into resorts of owner-occupied apartments.

The developers profited from the fact that the militias had largely superseded the authority of the state. In times of absent, weakened or corrupt state authorities, developers took advantage of the public domain of the coastline and illegally privatised access to the beaches. Originally the area of most of the beach resorts did not exceed two hectares, yet many doubled or even tripled their surface area when it became possible to do so (see figure 9.4).

The development of a new resort would frequently be under the patronage of a militia leader. Developers could thus disregard official planning and building laws without resistance. Thanks to the close relations of some of the militia leaders to the marginalised and confessionally segmented public authorities some developers could even obtain formal public approval for the illegal occupation of the public domain. Militia leaders profited from these projects as they received a financial contribution from the developers and took a tax from the apartment buyers. Furthermore, some militiamen found work as guards or watchmen in the new complexes.

The financial success of these resorts led other investors to copy this model.

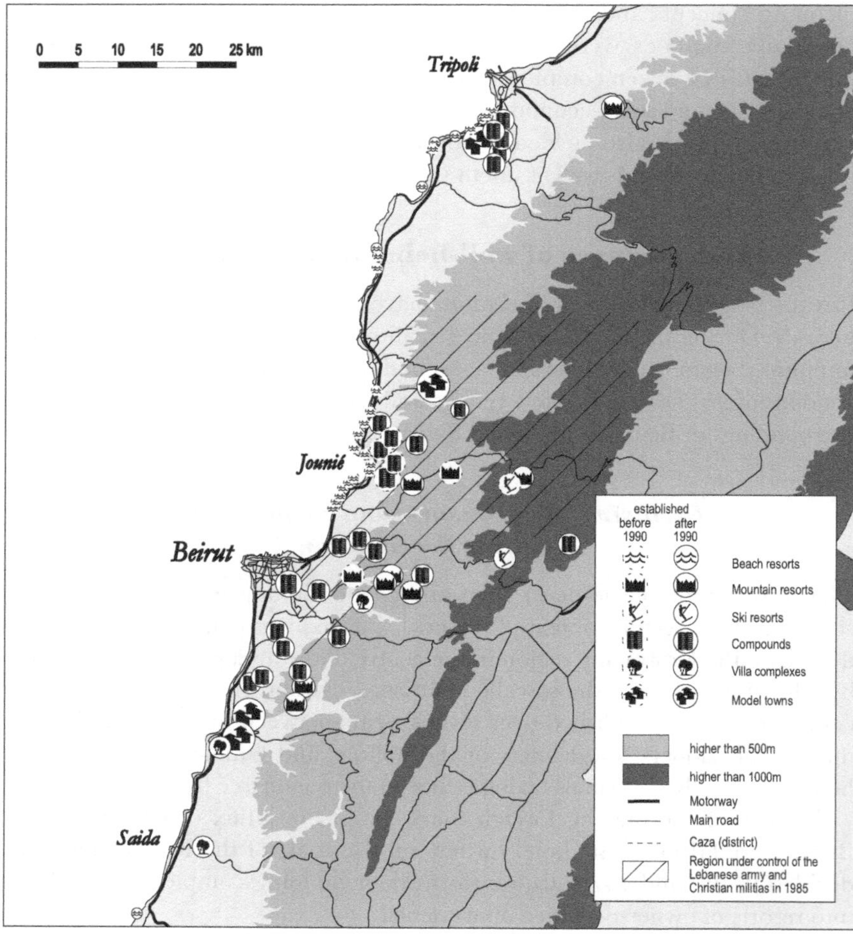

Figure 9.3 Guarded housing estates in Lebanon. Source: Glasze and Alkhayyal 2002

Indeed it proved to be very successful. Again and again, new 'rounds' of the civil war led to additional waves of families looking for a refuge in the region. Thanks to the economic boom of the 1960s and 1970s, as well as the influx of money from Lebanese guest workers in the Gulf states, there were a lot of families with enough money to buy a first if not a second apartment.

The first gated condominiums in Lebanon emerged in the mid-1980s, also in the Kesrouane region. The clients of these first condominiums were for the most part upper- and middle-class Christians who had left Beirut or Tripoli. Many had already lived in beach resorts where they took refuge and came to appreciate this kind of housing. Once they decided to stay in the region, many looked for larger apartments in a condominium estate.

People who moved to a gated housing estate during the war were in most cases attracted by the secure supply of water and electricity. During the war,

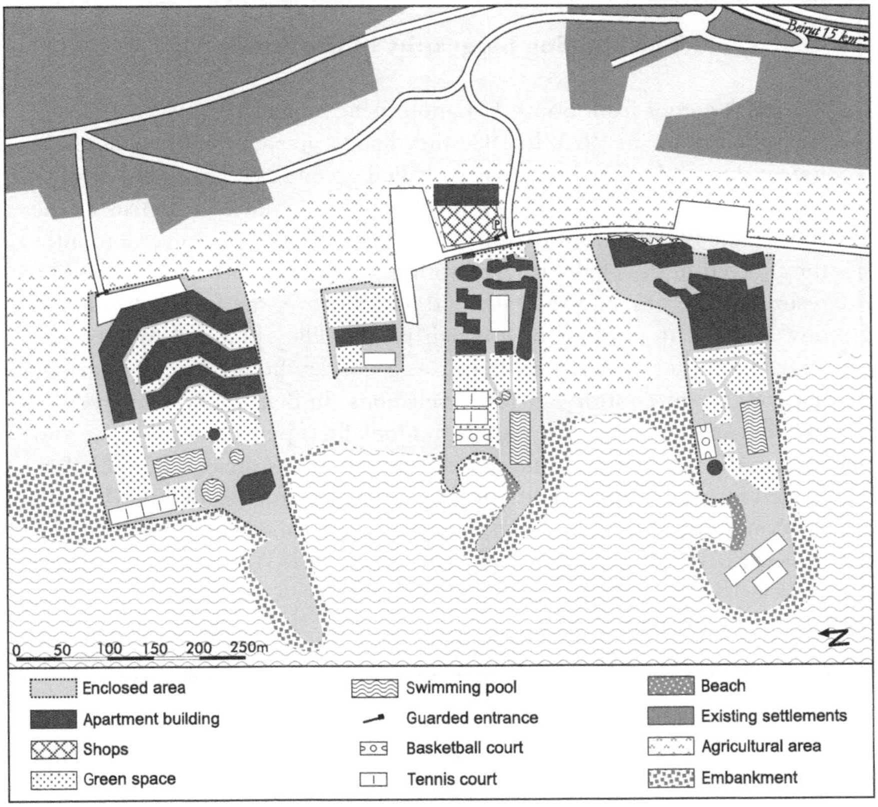

Figure 9.4 Three resorts in the Kesrouane region. Source: Glasze 2003

the resorts and the condominiums constituted enclaves where the secure supply of water and electricity was guaranteed, in a country where public services had more or less stopped functioning (Awada 1988; Dajani 1990). Generators and enormous water tanks or wells made the complexes largely self-sufficient. Families who had been driven out of their old homes in particular valued the protection offered by fences, guards and life in a closed community. During the war, the guarded residential complexes enabled a part of Lebanese society to keep a minimum quality of everyday life, which could rarely be found outside the gates.

Global lifestyle and property speculation – the post-war boom

In the post-war real-estate boom of the early 1990s more than 20 new projects – predominantly condominiums and gated settlements – were started in the suburban surroundings of Beirut and Tripoli. Contrary to the first guarded residential complexes in the 1980s those of the 1990s were established by

The housing biography of the family 'A'

Family A comes from South Lebanon, near Saida. They were expelled from their home in 1975. In 1982 they had to give up their new house in Beirut due to the fighting at the so-called green line. Together with her daughters, Mrs. A went to her brother's home, north of Beirut in the Kesrouane region far away from the chaos of Beirut. For several months they stayed in the chalet of her brother – a small apartment in a beach resort. As the situation in Beirut did not improve, the family decided to buy a chalet in another new beach resort. They planned to keep the apartment after the end of the war as a holiday home. But the war did not end; on the contrary, in 1985 their house in Beirut was destroyed. So they decided to stay in the region and look for a permanent home – and to keep the comfort of a guarded housing estate. At the end of 1985 they moved into a big apartment offering 250 square metres in a newly built condominium. Mrs A sums up their life in the condominium during the civil war: 'We did not suffer like the others did. We always had water and electricity. And we have been in a big house, a community . . . It has been a world apart.'

developers almost all of whom had experience in the real-estate sector. Most of them had become familiar with guarded complexes outside Lebanon. Some had been involved in planning, building or administering such complexes in South America, the USA or Saudi Arabia. After the end of the war they had the expectation that Lebanon would once again become an 'eldorado' of the regional real-estate sector. Their conviction was based on two assumptions. First, there was the prospect of a large unleashed demand for luxurious dwellings. Many of the interviewed developers assumed that there was a huge deficit in the housing supply especially for young urban couples. Furthermore, the 'millions of wealthy Lebanese' abroad, who would return to their mother country or at least buy a second residence there, would foster an explosive demand for housing. Second, many developers acted on the assumption that Lebanon could recover within a few years as the main economic and tourist centre of the Arab Middle East. Therefore capital investments in real estate seemed to promise high and rapid returns on investment – especially investments in luxury apartments.

In large-scale publicity campaigns the investors tried to reach their target group in Lebanon as well as the Lebanese abroad (Glasze 2003: 162–74): 'Far away and yet near to the city'. Consistently, texts and images represent the location of the developments as green and unsettled areas. Locations are described as 'unspoiled green hills' with an 'unobstructed view'. Environments are presented as healthy alternatives to city living: 'Get healthy and benefit from a real paradise'. So the complexes are differentiated from the city as a place of 'pollution', 'noise' and 'crowdedness'. Peripheral locations are presented as an advantage, but projects that came on the market at the end of the 1990s tried hard to let their location appear not too remote, advertising themselves as 'only a short distance from Beirut' and 'you will reach Beirut in a mere 15 minutes'. Maps now regularly show private schools and universities in the locality – a change in representation that seems to be the consequence of re-established competition with central Beirut as residential area.

This is also evidenced in the way amenities and common services are advertised. The 'self-sufficient' supply of water from wells and huge tanks and electricity from private generators is consistently described in advertising literature. Security and maintenance services 24 hours a day are mentioned in every one of the advertisements. The message is clear: in order to get a safe and comfortable life you should move to a private guarded neighbourhood. Private neighbourhoods face the competition of a more secure Beirut housing market by emphasising their comparative advantages. One well-established instrument of advertisement is the combination of apparently happy people with a product: those who consume this product will be happy and fulfilled too. Brochures and websites are filled with young and apparently happy people – often practising sport or other leisure activities. The advertisement seems to attempt to convey the values and ideals which it presumes are to be found within the target group of 'young Lebanese families'. Sport thereby becomes an element of social distinction.

The presentation of the social life in the complexes is inconsistent, however. On the one hand, several complexes are designated as 'communities', playing with the double sense of 'community' as 'settlement' and as 'social community' in the sense of *Gemeinschaft*. On the other hand, there are hardly any statements or pictures of communal activities (other than sports). This seems to be the consequence of the prejudices many Lebanese have about common housing – regularly associated with social housing and poverty. The absence of this aspect of private community life is in contrast to advertisements for US 'gated communities'. One slogan for a huge condominium complex says, 'what is shared are only the benefits'.

The promise of a better quality of life – good environmental conditions, leisure-time facilities and a 'modern', international and well-educated social milieu – has successfully appealed to young households of the Lebanese middle class. Returning Lebanese migrants, coming back to the country after the end of the war, chose guarded residential complexes as an orderly and well-governed place to live – drawing on expectations they had become accustomed to abroad. Both groups have been attracted by the image of a global lifestyle, constructed on Western models – as promised by the advertisements. In this manner, a young Lebanese man, who returned to Lebanon in the early 1990s after living in South and North America for several years, describes his life in the condominium Al Majal (figure 9.1): 'I almost feel like I were still in Miami here; there is a basketball court, there is a pool, there will be a highway to the city . . .'.

However, by the end of the 1990s it became clear that the expected demand was overestimated and fell below the supply of luxury residences, which came on the market in large numbers. Wealthy Arabs from the Gulf states only reluctantly bought real estate in Lebanon and the return of Lebanese from abroad almost came to an end with the lasting economic crisis since 1996. Since then the emigration of young Lebanese has exceeded repatriation. Moreover, on the back of the recession at the end of the 1990s many households of the Lebanese middle class were no longer able to finance a dwelling in a guarded neighbourhood. To cap it all, in the Beirut region, the guarded residential complexes which are mostly situated in the periphery are now faced with the new trend of 'city-centre living' – partially fostered by the more solid reconstruction of Beirut's city centre.

Comparing the motivation to move into a guarded complex with the period during the war it is significant to note that the 'secure supply of water and electricity' together with 'possibilities for children to play in a safe and green surrounding' were still the most important motives cited by people who moved to private guarded neighbourhoods (figure 9.5). Even after the reinstallation of public authority, the inhabitants are still attracted by the private provision of local public goods within the guarded housing estates.

On a micro-scale, the complexes fill the gap left by ineffective or insufficient public authority on a municipal and national scale. This is not only true for utilities but also for regulation. The bigger projects in particular attract customers by establishing regulations for zoning and building which are much more severe and better implemented than public regulations and serve to preserve and assure the aesthetic quality of the local environment. The developers of the 'private model town' Mechref, for example, established regulations for each section, which were far more rigid than the regulations of public planning for this area. They limited the areas for multiple family dwellings and prescribed a minimum lot size of 1,000 square metres, a maximum surface exploitation of 30 per cent, a maximum total exploitation of 40 per cent, a maximum height of buildings of 10.5 metres and the coverage of a minimum 60 per cent of roofs with red tiles. Unlike public planning rules, implementation of these prescriptions was ensured by ownership rights and by contract.

Segmented governance patterns – fragmented urbanism

In Lebanon the development of the first guarded residential complexes, in particular in the Christian Canton, was closely related to the specific context of the civil war and was driven by the demand for 'enclaves of well-being' in times of insecurity. Contrary to that, the second wave of development in the 1990s was more connected to supra-national trends and mainly driven by the supply side, speculative investment in particular. However, the analysis of the governance patterns in both phases shows that in both phases the actors proceeded in a complex social context, which in each case provided a rationale for

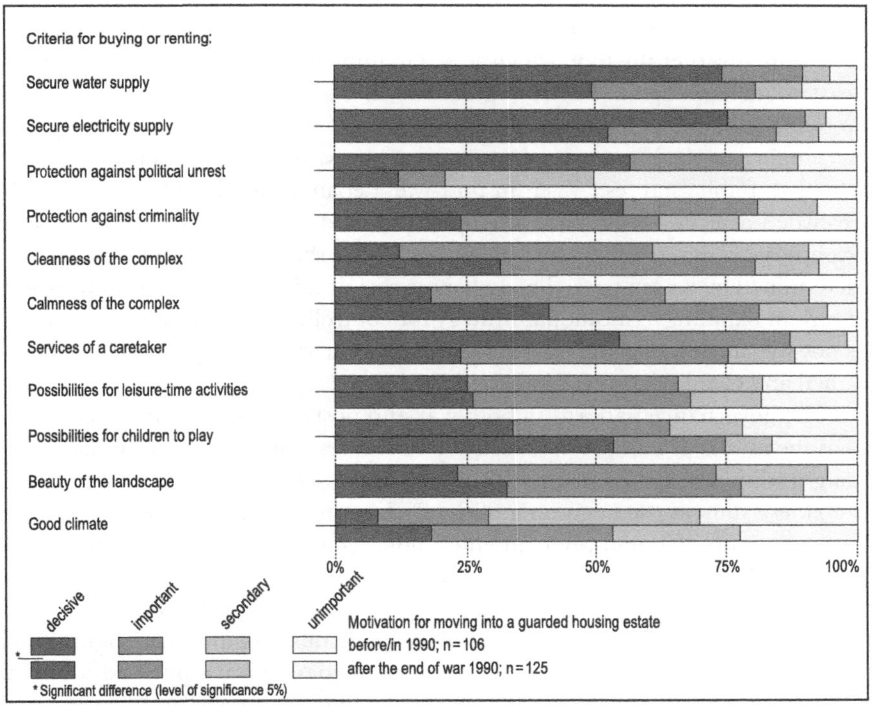

Figure 9.5 Motivation to move to a guarded housing estate. Source: Glasze and Alkhayyal 2002

the emergence of guarded residential complexes as an individually meaningful option.

After the war, the Lebanese state remained unable to secure a reliable supply of collective public goods. Many families distrust the efficiency of the state. The choice of dwelling in a guarded complex is a more reliable way to achieve secure and well-serviced urban living conditions. Why does the Lebanese state not fulfil core tasks of public central and municipal authorities? Leaving aside issues of taxation capacity, the answer partly lies in the institutionalisation of segmented interaction patterns. Even the re-established Lebanese state does not fulfil the idea of a unitary organisational unit, where public authorities have certain autonomy in relation to the particular interests of individual groups and orient their activities towards the public interest. Ministerial posts, in fact all posts in the administration, are occupied in equal parts by Christians and Muslims. Most employees in the public service owe their posts to confessionally determined preference by a patron. Therefore they retain a client connection with this patron. Also, after the end of the civil war, public representatives in the highest ranks of the state present themselves expressly as members and representatives of 'their' confessional group (Picard 1994). The ideal of public actors who initiate and moderate the construction of the public interest in public debates, and assure their adaptation, fails in the context of segmented interaction patterns. The state is rather the tool, which serves to implement particular interests or at least to defend them. Public regulations are interpreted by participants of another segment as an illegitimate use of the client structures and rarely find acceptance as legitimate implementation across all main dimensions of the (divided) public interest. Even in post-war Lebanon public policy is regularly regarded as a non-legitimate attempt to expand the power of one particular group. As a consequence, the public control of urban development tends to be determined by laissez-faire. The Lebanese sociologist Khalaf doubts in principle if 'urban planning as a rational strategy for controlling and allocating resources' is possible in view of a sectarian state organisation and the importance of confessional and client relations (Khalaf 1985: 214).

The governance patterns in both war and post-war periods have been such that the interaction between investors and leaders endorses a pattern that has a long tradition in Lebanon and dates back to the segmental construction of the Lebanese state at the beginning of the 20th century. The client entwinements between economic and political elites along confessional lines are complexly connected with liberal public regulation and an unreliable state provider. The power and resources behind public functions are segmented, following a more or less proportional representation of the religious communities (figure 9.6). Since 1943 all governments of independent Lebanon have pursued a liberal political course, granting the private sector almost unrestricted leeway, and limiting public regulations and supply to a minimum. Only against the background of these decades-old traditions can it be understood that nearly all of the interviewed developers and owners were convinced that the possession of a property gives the right to develop it. Furthermore, the absence of even a minimum of public regulation during the civil war often blurred the distinction between legality and illegality. The option to legalise nearly all offences against the town planning and building law, by paying a fee, confirmed the conviction of many developers that building laws should not be taken 'too seriously' in Lebanon. The attempt of the Hoss government to strengthen the supervision of construction activities in 1999 was interpreted by many developers as an unequalled and outrageous interference in their property rights. Ideas for the long-term protection and the sustainable creation of public goods seem doomed to failure. Therefore the concept of self-sufficient, guarded residential complexes finds a fertile environment.

French thinker Michel Foucault has developed the neologism *gouvernementalité* to describe the whole of the institutions and practices by means of which one steers human individuals – from administration to education (Foucault 1989: 97). This concept explicitly encompasses the idea that individual subjects are constituted and constitute themselves within specific discourses and power relations. Therefore the deeply rooted segmented organisation of the Lebanese society and state which stretches from the composition of political institutions (parliament, ministries, public administration) to questions of civil status and individual identities might be described as specific 'Lebanese governmentality'

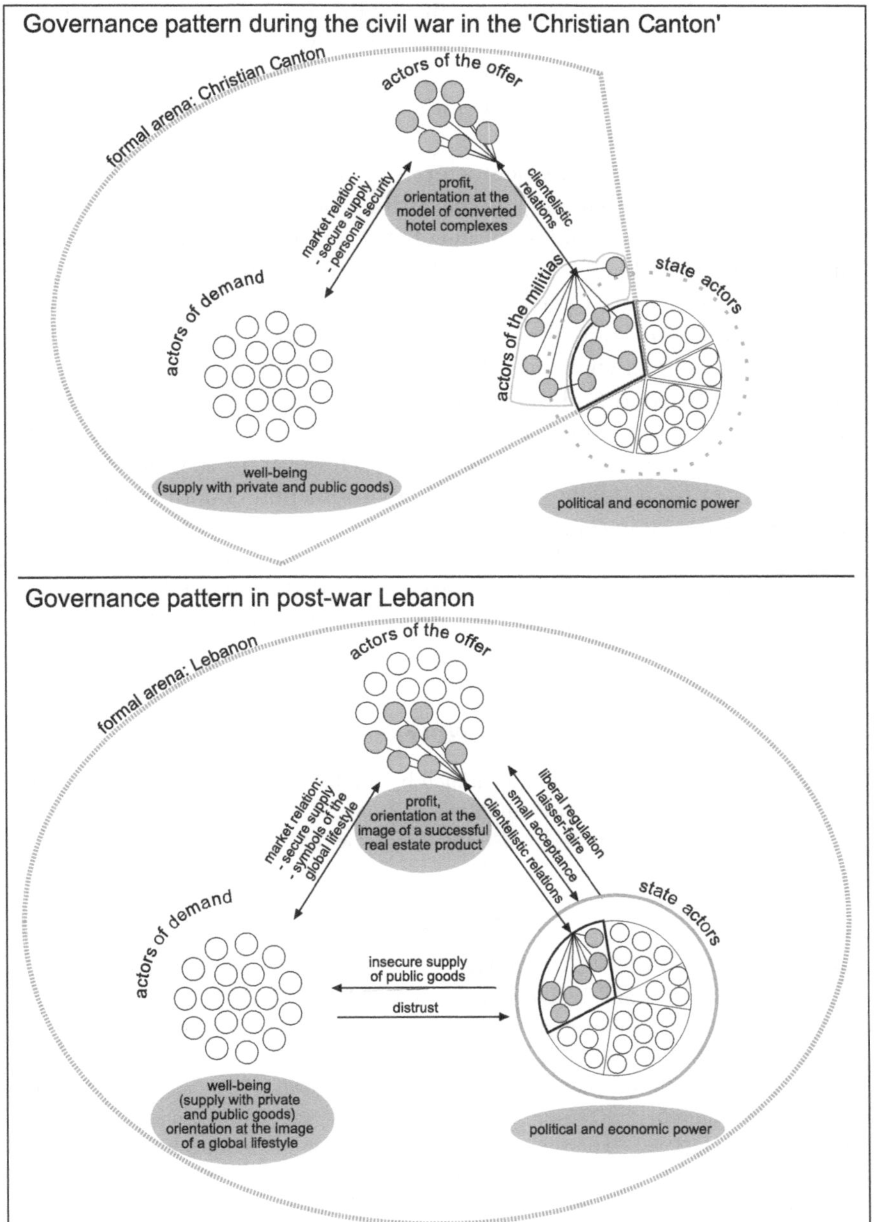

Figure 9.6 Urban governance in Lebanon. Source: Glasze and Alkhayyal 2002

which is often taken for granted as a quasi-natural structure. However, there are those who benefit from this structure and those who do not. The dominating elites of the different religious and confessional communities, especially, profited and continue to profit from the segmented structuring of social relations. It is not surprising that they kept or strengthened the segmented construction of the Lebanese state, by formalising the differences between the confessional groups and placing them as a basic condition prior to the construction of the state and by linking an anti-confessional rhetoric with a confessional practice (Mahdi 1996; Reinkowski 1997).

Having said this, there are new actors and interests emerging from the more stable middle-class Lebanon. The successful protest of Lebanese environmental groups against the building of new beach resorts in Jbeil indicates a new kind of alignment. They asked the Lebanese state to protect the public property of the coast against privatisation and to keep international agreements concerning the protection of monuments and the environment (Al-Azar 2000). They have, in effect, challenged the public authorities to act like real public authorities. The future will show if a reform of the Lebanese state can develop under such pressures – and whether the more mature society and economy is able to re-create the idea of a public interest. However, in view of a still unreliable supply of local public goods (bad environmental conditions, scarcity of green areas, uncertain supply of water and energy) and an urban development that remains hardly controlled, all of the interviewed investors, architects and developers were convinced that the private guarded residential complexes are the form of living for the future. The plans to realise them are already in their drawers.

Notes

1 The fieldwork for this study was funded by the Deutsche Forschungsgemeinschaft (DFG – German Research Foundation).

2 Young refers to Hourani's (1976) essay 'Ideologies of the mountain and city'.

Bibliography

Abu-Lughod, Janet L. (1987) 'The Islamic city: Historic myth, Islamic essence, and contemporary relevance', *International Journal of Middle East Studies*, 19: 155–76.

Al-Azar, M. (2000) 'Jbeil protests at Marina Plan for "last public sandy beach"', *Daily Star* (Beirut), 9 May 2000: 17.

Awada, Fouad (1988) *La Gestion des services urbains à Beyrouth pendant la guerre*, Paris: Pratiques Urbaines 5.

Dajani, Nabil H. (1990) 'The role of urban institutions in the management of daily conflicts in Beirut', unpublished conference paper, Les Espaces de l'orient mediterranéan conference, Université de Lyon 2, November 1990.

Davie, M.F. (1994) 'Guerres, idéologies et territoires: L'Urbanisation récente de la côte libanaise entre Jbayl et Sayda', *Annales de géographie*, 575: 57–73.

Foucault, M. (1989) 'La Gouvernementalite', *Le Magazine littéraire*, 269: 96–103.

Glasze, G. (2003) *Die fragmentierte Stadt. Ursachen und Folgen bewachter Wohnkomplexe im Libanon*, Opladen: Leske and Budrich.

Glasze, G. and Alkhayyal, A. (2002) 'Gated housing estates in the Arab World: Case studies in Lebanon and Riyadh, Saudi Arabia', *Environment and Planning B*, 29 (3): 321–36.

Goodwin, M. and Painter, J. (1996) 'Local governance, the crises of Fordism and the changing geographies of regulation', *Transactions of the Institute of British Geographers*, 21 (4): 635–48.

Hourani, A. (1976) 'Ideologies of the mountain and city', in R. Owen (ed.) *Essays on the Crisis in Lebanon*, London: Ithaca Press: 33–42.

Khalaf, S. (1985) 'Social structure and urban planning in Lebanon', in E. Mauer (ed.) *Property, Social Structure and Law in the Modern Middle East*, New York: State University of New York Press: 213–35.

Le Galès, P. (1995) 'Du Gouvernement des villes à la gouvernance urbaine', *Revue française de science politique*, 45 (1): 57–95.

Mahdi, A. (1996) *L'Etat confessionelle. Le Cas libanais*, Montreuil: La Brèche.

Pahl, R.E. (1970) *Whose City? And Other Essays on Sociology and Planning*, London: Longman.

Picard, E. (1994) 'Les Habits neufs du communautarisme libanaise', *Cultures et conflits*, 25 (4): 49–70.

Pierre, J. (1999) 'Models of urban governance – the institutional dimension of urban politics', *Urban Affairs Review*, 34 (3): 372–96.

Raymond, André (1989) 'Espaces publiques et espaces privés dans les villes arabes traditionelles', *Maghreb, Machrek* 123: 194–201.

—— (1994) 'Islamic city, Arab city: Orientalist myths and recent views', *British Journal of Middle Eastern Studies* 21: 3–18.

Reinkowski, M. (1997) 'National identity in Lebanon since 1990', *Orient*, 38 (3): 493–515.

Tuéni, H.E.G. (1998) 'From the geography of fear to a geography of hope', in P.G. Rowe and H. Sarkis (eds) *Projecting Beirut: Episodes in the Construction and Reconstruction of a Modern City*, München, London, New York: Prestel: 285–96.

Wirth, Eugen (2000) *Die orientalische Stadt im islamischen Vorderasien und Nordafrika; Städtische Bausubstanz und räumliche Ordnung, Wirtschaftsleben und soziale Organisation (Band 1)*, Mainz: Philipp von Zabern.

Young, M.D. (1995) 'Cities of salt', unpublished manuscript, Beirut.

10

THE PURPLE JADE VILLAS (BEIJING)

A golden ghetto in red China

Guillaume Giroir

Abstract

This chapter presents the results of a field study conducted in June to July 2000 in a gated community (*bieshu qu*) near Beijing, the Purple Jade Villas (*Ziyu shanzhuang bieshu*, literally, the 'Villas from the Mountain Village of Purple Jade'). This private residential complex of 400 luxury villas occupies a total area of 66.6 hectares and is situated in the north of the capital, 12 kilometres from the centre, between the fourth and the fifth ring roads.

At present, no gated community has been described in China, because it is a new phenomenon born in the 1990s, and especially because this is a taboo micro-territory in Communist China. This study is the first of a series of empirical and theoretical works relating to this kind of post-communist micro-territory in China (Giroir 2002, 2003, 2004a, 2004b).

In this chapter, this new phenomenon has been analysed at two complementary levels. First, this case study reveals several similar elements with other upmarket gated communities described elsewhere, but also many particularities resulting from the transition of China between socialism and capitalism.

This study examines the various constituents of this luxurious place: situation, architecture, environment aspects, prices, nature of residents, leisure facilities and recreational practices, and the role of the club house as central place of this golden ghetto. It is also shown that the marketing strategy of property developers contributed very much to create the illusion of a space totally different from Beijing. So, in a classic approach, the Purple Jade Villas can be interpreted as an extreme segregative phenomenon insofar as the luxury is completely disproportionate to the standard of living in Beijing.

But this residential enclave can also be interpreted as a transitional phenomenon. It highlights the dualism of the Chinese property market, the dubious relations between money and power, the remaining public land rights and the self-segregation of the Chinese elite as a residential response to an ideological constraint.

Second, this chapter presents at another scale of analysis the ambivalent and conflictual relationships between this luxury enclave, the metropolitan area and globalisation. In many respects (dollarisation, European taste and international openness), this luxurious enclave appears like a privileged, isolated and globalised oasis. But in other respects (insecurity risk, polluted environment, urban spreading) it closely depends on the Beijing megapolis and cannot escape to its externalities.

Introduction

After the end of China's egalitarianist Maoist period, a point marked by the 1978 reforms, a process of socio-spatial differentiation started on every scale – inter- and intra-regional, between towns and regions and within urban spaces. The recent phenomenon of luxury villa areas on the outskirts of large cities makes such disparities tangible. These peri-metropolitan micro-territories, which have been little investigated, reveal the extreme form that economic disparities may take in China. They bear clear similarities to the gated communities found in the United States and elsewhere and sit strangely alongside the pre-reform Chinese urban landscape.

In this chapter I describe the case of the Purple Jade Villas (*Ziyu shanzhuang bieshu* – literally, the 'Villas from the Mountain Village of Purple Jade'). This is one of the earliest examples of an upmarket private residential estate in Beijing and now lies in the inner ring of the city's fast expanding modern belt of commodity housing. The Purple Jade Villas case is interesting because it provides an insight into the way market-led urban development is proceeding apace in the People's Republic of China, which still claims to be socialist. In the commentary in this chapter I concentrate on the discontinuity between this kind of development and what existed before. Webster, Wu and Zhao, in their chapter, look at the continuity.

In order to appreciate just how great a change is being wrought in the spatial–social ecology of Chinese cities it is first appropriate to show the particularities of private spaces like Purple Jade Villas. I highlight the distinct features of this luxury haven and consider them against the backdrop of the wider social and economic landscape of the Beijing metropolis. I draw attention to the strategies of property developers, who emphasise design and scenery in placing Purple Jade Villas as a distinctively new kind of space in Beijing. Second, I consider how the distinct features of this development demonstrate China's openness to globalisation. I also comment upon the limits of this attempt to disconnect from the roots of the host metropolis.

A luxury haven at the gates of Beijing

Luxury and its socio-economic context

Luxury and metropolitan situation

The location of Purple Jade's luxury villas area responds to the dual demands of its rich tenants. At 12 kilometres north of Tian'anmen Square, between the fourth and the fifth ring roads, it is near enough to the city centre to make commuting bearable but also 15 minutes' drive from the international airport of Beijing. It is part of a wider phenomenon of the upper class escaping towards the outskirts, away from Beijing's intense pollution, traffic jams and other central city externalities. A comparison with Paris highlights some of the costs of central city living that are driving suburbanisation: there are only six square metres of green space per inhabitant in Beijing, compared to 11.6 square metres in Paris; particles suspended in the air (dust in particular) reach 319 milligrams per square metre compared to 34 milligrams per square metre; the concentration of sulphur dioxide is 135 milligrams per square metre compared to 30.8 milligrams per square metre for Paris (Yang 2000: 229).

Ways of development: dubious relations between money and power

The Purple Jade Villas case provides an indication of the way such villa areas are developed in China. This huge property investment was the result of an association, ten years ago, between Mrs Liu Xiaoqing and Huang Ziyu. The former is from Sichuan Province and made her money by becoming a television and movie star in Hong Kong. The latter is now a member of the Political Consulting Conference of the Chinese People and succeeded in obtaining – thanks to her relations with high leaders of the government – the necessary administrative authorisations for a politically sensitive estate transaction. This association between the sphere of money and power ended recently when Mrs Liu sold Mrs Huang her shares. Mrs Huang, the sole owner, has lent her first name, Ziyu ('purple jade'), to this housing programme. Mrs Liu is now a billionaire and a shareholder in many luxury villa areas around large Chinese cities. She was tried on various counts of tax evasion in the People's Republic of China and the case was given wide coverage by the press. Famous for performing historical figures (notably Empress Dowager Cixi), she has become the 45th richest person in China according to *Forbes* magazine. Charged with tax evasion, she had to pay 5.7 million yuan and was sentenced in June 2000. The Court sequestered 18 of her luxury villas.

Out-of-range price level

The Purple Jade Villas area will eventually comprise 400 villas built in three successive stages (figure 10.1). The conception of the villas was subject to advanced market research, hence their subtle differentiation into 20 or so

Figure 10.1 Purple Jade Villas (Beijing): a European-styled luxury villa. Photo: Guillaume Giroir 2000

models. They vary according to their surface area, architectural style and home design and resolutely escape from the uniformity of the city's existing housing estates.

The sale price level of the villas makes them the province of the wealthy (see table 10.1 overleaf). Nearly 80 per cent of the residents are Chinese (or of Chinese origin), coming equally from Beijing and the rest of China. A minority are from Hong Kong and foreign countries. In total, this housing community is currently thought to have 426 of Beijing's richest citizens (figure 10.2 on p. 147).

The purchasers have title to their property in the form of a 70-year ground lease. The leasehold can be extended after this period – a measure which maintains the fiction of the ideological continuity of the regime and, in particular, the collective aspect of land in China. About 20 per cent of the villas are rented – in all cases the rent being paid by large firms and administrations. Ordinary Beijing professional people could never afford to pay rents of US$6,800 a month for a 298-square-metre villa or US$7,500 for the 343-square-metre model. To this sale or rental price are added high service charges to cover intensive security and park maintenance. Ordinary residents pay on average US$1.5 per square metre per month, that is to say, for 343 square metres, a sum close to US$515.

Table 10.1 Surface and sale price of the Purple Jade Villas

Type of villa	*Built surface area (m^2)*	*Floor surface area (m^2)*	*Price/m^2 (US$)*	*Total price (US$)*
I	343.63	632.22	3,039	1,144,288
II	298.70	481.17	2,983	890,126
III	283.28	426.19	3,235	913,158
IV	242.10	492.17	3,106	751,992
V	183.40	160.56	2,514	461,916
VI	168.60	155.73	2,343	395,000

Source: presentation booklet of the villas (prices as of July 2000)

Demonstrating and showcasing luxury

Leisure facilities and recreation

Leisure facilities are a major feature of this luxury villa and reflect the high standard of living of its residents and their style of business. Recreation is closely linked to business life – here as anywhere else, big contracts or decisions are taken in secret and exclusive places. The retired and semi- or early-retired are well represented and this contributes to the setting up of a wide range of sophisticated leisure facilities. For the foreign expatriate executives, everything is done to try to soften the feeling of exile. Many of them stay in Beijing with their families for long periods of time. A few villas have private swimming pools, which supplement the two communal outdoor swimming pools. One of reduced size turns into an ice skating rink in winter, thanks to the harshness of the Beijing climate. The other, a 50-metre pool, stages from time to time various aquatic shows performed by Chinese or foreign troupes. There are four tennis courts at the disposal of residents and a golf course is being built.

Showcasing luxury and marketing strategy

Purple Jade Villas occupies a total area of 660,000 square metres, mainly consisting today of a villa district of 130,000 square metres and 'natural' spaces of 223,000 square metres. Compared to the other luxury villa areas around Beijing, Purple Jade Villas is remarkable given the exceptional profusion of its park land and lakeside areas. It is a startlingly green area. According to the sales pitch, the villa area, which is part of the green belt of Beijing, claims to be something of a model in matters of environment. Two 30,000-square-metre artificial lakes, both excavated, are stocked with Wuchang fish and carps for angling. The complex has a central grand lawn of 30,000 square metres. In the middle is a big arbour made of canvas with white tables and chairs for receptions and garden parties. The entire 'garden suburb' is also protected by a 133,000-square-metre range of windbreak trees facing north-west, aimed at reducing the nuisance of sandstorms and dust coming from Xinjiang and Mongolia.

Figure 10.2 Rich Chinese angling in an artificial lake. In the background, new emerging high-rise blocks in urban smog. Photo: Guillaume Giroir, 2000

Several thousand trees were planted where an old village used to be. Farm production remains only under the residual and idealised form of a few peach, cherry and pear trees chosen as much for their quality taste as for the beauty of their blossoms. Likewise, some parts of the park show a profusion of flowering borders. The African marigolds (*yueji*), emblem of Beijing, flower from March to October – symbolic jewels in the crown for the luxury villas and their inhabitants.

The environmental niching led the developers of Purple Jade Villas to collaborate with the Chinese Wild Animal Association and the Beijing Bird Breeding and Bird Loving Association (*Beijing ainiao yangniao xiehui*). The park is consequently something of a zoological garden for specialist (luxury) breeds. The remarkable presence of all these birds in a Beijing housing estate is reminiscent of the peacocks and other exotic species that once graced the grounds of the aristocratic houses of Europe (figure 10.3 overleaf). It is a symbol of both aestheticism and elitism in line with the image of these luxury villas. More pragmatically, it is a compensating factor enhancing the value of this scheme above the ambient value for this location in Beijing and compensating residents for nearby nuisance and urban chaos. The 'village of the mountain' evokes pure air and a protected environment, and contributes to the illusion of a place free from metropolitan contingencies.

Figure 10.3 The peacock, symbol of luxury. Photo: Guillaume Giroir, 2000

The clubhouse – central place of the golden ghetto

The clubhouse is the principal and central element of this residential enclave, being built in the middle of the lake where it is at the geometric centre (figure 10.4). With the modernity of its big reflective glass surface and the verticality of its three floors, it represents, on the architectural level, the most visible emblematic sign of the villa area. The range of indoor activities it offers makes it the principal meeting place. Social screening for access to the clubhouse is ensured by an annual membership fee of US$1,800 per person plus US$70 per month. Casual visitors must pay US$100 a day. For this price, the clubhouse offers a large range of activities including indoor swimming pool, fitness room, cinema, coffee shop, tea room, restaurant and conference hall.

An abstracted space?

The commercial strategy of the Purple Jade Villas' developers consists in stressing the villas' otherness compared to the rest of the metropolis. The luxury of Purple Jade Villas arises from its material quality and its diverse and luxury services, but also from its exotic features – even its antithetical relationship with the local environment. It puts at the disposal of residents what common Beijing people are deprived of: space, parks and leisure facilities. In many respects Purple Jade Villas is a kind of metropolitan offshore area. The

Figure 10.4 The club house, central place of the golden ghetto. Photo: Guillaume Giroir, 2000

developers have endeavoured to model the areas on similar spaces throughout the world and tried – successfully – to produce a space in which one can forget the immediate locational context of Beijing and its suburbs. Thus Purple Jade Villas is a luxury haven among so many others in a globalised world.

Purple Jade Villas or the impossible globalisation

Demonstrations of globalisation: dollarisation, European taste, international openness

The Purple Jade Villas economy relies essentially on the US dollar. The membership fee for the club house and the purchase and rental payments themselves are paid in dollars. Everyday spending is made in yuan, but the amount varies according to dollar–yuan parity. This relative dollarisation contributes to the feel of an extraterritorial enclave.

If, on the monetary level, westernisation goes with Americanisation, on the landscape and architectural level the villas are far more marked by a European influence. Actually the developers' presentation booklet strongly emphasises the European style of balcony and garden fence, and imported materials and equipment are central to the scheme. If much of the furniture is made in Hong Kong, ceramics, sanitary fittings, sofas and sporting facilities are from Italy while household appliances are German. These imports are an integral part of the sales argument.

Purple Jade Villas' openness to global consumption culture is hardwired in its TV services – CNN International, Deutsche Welle, TV5 Asia and so on. The reading room in the clubhouse is laden with international magazines about stars, fashion, design and the business world.

Limits to openness

A more subtle analysis of this micro territory leads to a more considered reflection of the globally focused openness of the enclave.

Food: open to the world or Hongkongisation?

The food offered at the clubhouse restaurant is a further indication of the villas' embeddedness into the globalised world of luxury. Among the 29 meals available at the self-service counter at the time of visiting there is almost no trace of Beijing food – one of the best Chinese regional cuisines. It seems to be a purposeful rejection of the local. By contrast, non-Chinese meals were clearly over-represented, with 11 different offerings. The niching strategy of the developers and management clearly extends to menus. However, southern Chinese food, especially from Guangdong Province (14 dishes), is the clear winner and one might read into this the spreading of an economic and cultural model from the southern seaboard provinces.

Obsession with security and forced integration into the local territory

In the realm of security, this globalising community is forced to face up to the reality of its contextual integration into a local territory. The wealth of residents and the value of their homes and possessions result in obsessive security. Each house is equipped with an alarm, while direct phone lines put residents in touch with a security service capable of intervening in emergencies. Entry to the villa area, a physically enclosed world separated from the city by a continuous wall, is protected by a warden. Access is reserved for occupants of the villas and their relatives and friends. Even the park area is dotted with imposing guardrooms. The private security officers are dressed as perfectly as soldiers. In addition to the fixed observation posts, soldier-watchmen constantly patrol, some accompanied by Alsatian dogs. Watched night and day, all year long by about 40 wardens, the Purple Jade Villas is a truly fortified place – a quasi-militarised space with its own quietly threatening sense of order and peace. Eden cannot escape a form of local tyranny when it is surrounded by chaos.

A deceptive model in matters of environment

One of the principal sales arguments by which Purple Jade Villas claims to be a model in matters of the environment seems to be something of a deception.

According to the property developers, the air and water quality of the housing area has been certified by the Beijing Central Laboratory of Research in the Environment. The presentation booklet of Purple Jade Villas (undated, unpaginated) claims that air and water are purer and clearer than anywhere else in Beijing. The idea of a micro-climate within such a limited territory can hardly be taken seriously. Despite the profusion of green and lakeside spaces, Purple Jade Villas does not escape from the pervasive urban smog that hangs above Beijing. The city's atmosphere is not intimidated by the dog patrols, guardhouse and walls. On the other hand, while the city faces a serious water shortage requiring the authorities to make dramatic appeals for rationing, Purple Jade Villas' vast surface of meadows, lakes and swimming pools owes its existence to an unremitting over-consumption of water.

Broken dream of autarky

Despite all the services offered to villa occupants, the dream of autarky has failed. Children have to go to the schools of surrounding districts. Residents go shopping in hypermarkets built near this high-purchase-power area. The biggest threat to the autonomisation strategy of the Purple Jade Villas has come from other citizens in search of their own less grand versions of Eden. Abutting the villa area are now a mass of commodity housing complexes consisting of monumental 20- to 30-floor buildings. The pressure of general urban sprawl pushing Beijing's suburbs outwards beyond the fifth ring road is compounded by the 2008 Olympic Games infrastructure. Some of the sites for the events are in this same sector and green land once at the urban fringe has soared in value and much of it has been developed at very high density. Purple Jade Villas has already become an enclaved luxury green reserve surrounded by many towers with a far less exclusive population.

Conclusion

The story of Purple Jade Villas is an extreme part of the bigger story of socio-spatial segregation in the metropolitan spaces of Beijing. Within the space of just a few years a Chinese urban society fast differentiating itself by income has also differentiated itself by urban ecology. Fast-growing China contains some of the greatest inequality in the world. The emergence of luxury in a country still officially communist is peculiar to the outsider. It would seem the most tangible of demonstrations of the gap existing between the views of the elite advocating equality and their practice and lifestyle. On the other hand, Purple Jade Villas shows how complex the transition process in China is. In China, few people have a problem understanding how communist rhetoric and luxury can coexist. Two systems are being permitted to coexist – to compete, even. More important than the substance of the continuing rhetoric, arguably, is the stability it gives. And stability is what the markets for luxury and all other levels of commodity need to thrive. And thriving it is.

Bibliography

Fraser, D. (2000) 'Inventing oasis: Luxury housing advertising and reconfiguring domestic space in Shanghai', in D. Davis (ed.) *The Consumer Revolution in Urban China*, Berkeley and Los Angeles, CA: University of California Press: 25–53.

Giroir, G. (2002) 'Le phénomène des *gated communities* à Pékin, ou les nouvelles cités interdites' (The phenomenon of gated communities in Beijing, or the new forbidden cities), *Bulletin de l'Association de géographes français*, 4: 423–36.

—— (2003) 'Gated communities, clubs in a club system. The case of Beijing (China)', proceedings of international conference in Glasgow on gated communities: Building Social Division or Safer Communities?, Glasgow, September 2003. Online. Available HTTP: http//www.gla.ac.uk/departments/urbanstudies/gated/gatedpaps/gatedconfpaps.html.

—— (2004a) 'The Fontainebleau Villas (Shanghai), a globalized golden ghetto in a Chinese garden', Proceedings of the International Annual Conference of the Royal Geographical Society, London, September 2003. Forthcoming in F. Wu (ed.) *Globalization and the Chinese Cities*, London: RoutledgeCurzon.

—— (2004b) 'The Yosemite Villas (Beijing), an American-style gated community, mirror of the representations of the rich and of emerging capitalism in China', Proceedings of the World Forum on China Studies, Shanghai, August 20–21. Forthcoming in Shanghai Municipal People's Government/Shanghai Academy of Social Sciences.

Yang, K. (2000) *Chixu shoudu* ('Sustainable development'), Guangzhou: Guangdong jiaoyu chubanshe.

11

CHINA'S MODERN GATED CITIES

Chris Webster, Fulong Wu and Yanjing Zhao

Abstract

The thesis developed in this chapter is that condominium – *co-rule* – is an institution deeply embedded in contemporary Chinese society and has survived the end of the centrally planned era in a number of intriguing forms. It has more than survived. While the condominium institutions underlying some modern developments are historical and in tension with market-led resource allocation, others have been created by the market. Still others are hybrid, lying somewhere along an evolutionary path and constituting an urban institutional experiment on a grand scale. This makes for a unique set of starting conditions for the future spatial shape of urban China. It also presents a unique set of institutional constraints on the evolving internal urban structure of cities. What Friedman and Weaver (1979) a quarter of a century ago termed selective territorial enclosure is, it seems, far from irrelevant to modern China's capitalist-led urban development.

Introduction

In 1979 John Friedman and Clyde Weaver published an influential little book in which they proposed the idea of *agropolitan development*. Consciously influenced by the Chinese model of development, this utopian and anti-free-market notion embraced three principles: selective territorial enclosure; communalisation of productive wealth, particularly land and water; and equalisation of access to the basis of the accumulation of social power (Friedman and Weaver 1979: 194–5). The third of these was the most utopian and could never have been expected to be achieved in China or anywhere else. The first two, however, are quite consistent with market-driven economic growth and spatial development – at least at the urban scale.

There are very good reasons why entrepreneurial firms and competitive government agencies acting like entrepreneurial firms might want to secure *selective territorial enclosure* and *communal forms of ownership*. One of the many intriguing features of modern Chinese cities is the juxtaposition and interactive evolution of various forms of condominium – literally 'co-rule' or communal ownership. An unexpected by-product of the centrally planned era is a ubiquitous and deeply embedded residual institutional framework of condominium. In fact, as we show in this chapter, there are many variants of condominium that have survived the socialist era and, strengthened by an intriguing nexus of market incentives, are shaping China's new mixed-market economy. Selective territorial closure and communal ownership, it seems, are far from irrelevant to modern China's urban development.

Our objective in this chapter is to reflect on some of the surprises and apparent contradictions found in urban China: the enclosure of premium locations to form private high-end gated communities, the resilience of enclosed former work unit residential neighbourhoods and their evolutionary path towards privately managed middle-income neighbourhoods, the enclosure of former rural collective territories to create vast privately managed and private–public partnership mixed-income suburban townships, the development of densely urbanised ex-urban villages within communally owned former agricultural collective territories, and the enclosure of entire sub-regions to create new metropolises. The condominium institutions underlying some of these developments are historical and in tension with market-led resource allocation. Others have been created by the market. Still others are hybrid, lying somewhere along an evolutionary path and constituting an urban institutional experiment on a grand scale. This makes for a unique set of starting conditions for the future spatial shape of urban China. It also presents a unique set of institutional constraints on the evolving internal urban structure of cities.

China's 'gated' cities within cities are created by various models of property ownership which confer relatively exclusive shared access by clearly delineating legal and conventional rights. Most are also delineated spatially by physical barriers. In contrast to the novelty of walls and gates in Western democratic societies, these outward signs of territory have existed for thousands of years in Chinese cities and are regarded by most as an unremarkable feature of the country's modern urbanisation.

Waves of condominium

From the start of the centrally planned era, productive resources in China came to be organised through industrial and institutional work units and agricultural collectives. Cities were tessellations of many quasi-autonomous work unit territories, or so-called *dan wei*, funded by and returning revenue to their supervisory government departments (Zhang 2002; Wu 2005). They provided housing, education and other services for employees, who by and large were not free to take their labour elsewhere. Work units tended to function as closed-membership joint production–consumption entities. Although the original intention of creating work units was political, i.e. consolidating grassroots governance, the actual function of their operation is not very different from that of economic clubs – entities that organise the supply of collective consumption goods to a set of members or organise co-production on the basis of members' shared productive assets (see Buchanan 1965 and Cornes and Sander 1996 for seminal developments of the economic theory of clubs; and Webster 2002; Webster and Lai 2003: chapter 6; and Heikkila 1996 for examples of club theoretic analysis of cities). Each *dan wei* competed with others for resources allocated by the supervising government department and had de facto control of resources it managed to secure.

Under this organisationally and spatially decentralised system, journeys to work, which were generally by foot or bicycle, were short, with housing and employment not untypically located on the same site. During the 1980s state enterprises wanting to improve the standard of employee housing built large amounts of new housing – in single work unit estates or organised jointly by several units into comprehensively developed 'residential districts' (*ju zhu qu*). The effect was to disperse homes further from the workplace, but not by great distances. This was a rationalisation of the co-owned housing that emerged in the early phase of the state-planning era. The organisation of supply was rationalised in this phase, for example, by several government ministries developing a housing estate together. So too was the spatial distribution of housing; for example, teachers' housing moved from school sites to large estates serving schools in a wider neighbourhood. The result was to add to the city an additional layer in the patchwork of work unit joint consumption clubs, this time not necessarily spatially contiguous with the production clubs organised by the same units.

In the interstices of the work unit city, less formal communal institutions survived and developed in traditional housing areas, such as the conventions governing the allocation of resources in neighbourhoods of courtyard housing. Traditional housing became the homes of workers in units too small or too poor to organise their own neighbourhoods. Housing was deemed an 'unproductive' sector and with most strategic investment directed to new industrial areas on the periphery of the city, the old urban areas were left intact and became progressively crowded and degraded. Courtyards originally designed for a single family became the home of two or three. During the Cultural Revolution anything up to 20 families were moved into the *hutongs* in Beijing. The courtyards were filled with low-quality structures, including single rooms accommodating entire families, with communal kitchens, toilets and wash areas and the once spacious homes became micro-neighbourhoods with their own informal rules for organising crowded communal life.

Overlaying the informal governance of subdivided courtyards and the formal governance in the *dan wei*, the *residents' committee* became the most local of government-organised mechanisms for collective action. These were set up as 'self-organised mass organisations', according to the 'Rules on the Organisation of Residents' Committees' promulgated in 1954 by the National People's Congress. They were (and still are) elected by residents and are meant to function as an intermediary between residents and higher levels of government and the latters' agencies. Typically, a residents' committee is in charge of 100 to 600 households and staffed by 7 to 17 people. Above these, the *street office* (*jiedao banshichu*) manages wider neighbourhood affairs, commonly for a population of 20,000 to 30,000. Like the residents' committee, the *jiedao banshichu* is strictly speaking an agent of local government (district or municipality) – both are state-initiated organisations designed to foster and facilitate collective action and control (Wu 2002).

In the transitional economy era, starting from constitutional reforms in 1982, new condominium institutions emerged or adapted from those existing. Urban land was decreed to be state owned and rural land to be communally owned by the villagers who had been organised into collectives since the 1950s. In the 1990s land-use rights were further decentralised and delineated. First, the local government (municipality or county) become the legal owners of state land. State enterprises, as de facto owners, were allowed to retain a share of the value of land transactions but the municipality monopolised the primary land ownership right. This form of land market, copied from Hong Kong, is effectively a land-leasing system. Foreign firms as external investors are almost exclusively restricted to this marketised land market, while in many cities a dual land system exists for domestic uses. In recent years, the land-leasing market in cities such as Shanghai and Beijing has been extended to include most domestic property developments. Once the right has been obtained from the municipality through conveyance, it can be further transacted in the secondary market through 'paid transfer of land-use rights'. The result has been the rapid development of land and property markets. In the cities, work units received partial or complete financial and administrative autonomy (Zhang 2002) and became de facto 'owners' of land formerly allocated to them administratively. Local governments similarly inherited alienable land-use rights over the land they occupied and were able to raise and retain their own tax revenues as well as lease out and develop land. The result of this mammoth and catastrophic decentralisation of land, administrative and fiscal rights has been cities of semi-autonomous entrepreneurial agencies, each one seeking to maximise the value of its territorial assets – principally land.

At a metropolitan scale, special economic zones (SEZs) acquired (via compensatory payments) and combined land-use rights from work units, local governments and village collectives. The consolidated land in this case formed the financial and physical resource base for the collective activity of city-building. Shenzhen SEZ, immediately to the north of Hong Kong, started purchasing land from rural collectives in 1980 and leasing it to private enterprises, including manufacturing and commercial firms and housing developers. By 2000 the city's population was approximately 7 million, roughly the size of Hong Kong, having grown from less than 40,000 in the early 1980s. Shenzen's GDP increased from 270 million yuan (US$31 million) in the early 1980s to 167 billion yuan in 2000 at an annual growth rate of over 30 per cent. The city's growth by the purchase of land-use rights, the enhancement of their value by infrastructural investment and their subsequent sale by leasing are an example of condominiumisation at a large scale. Under the enabling legislation that set the framework for special economic zones, dispersed rights held by villages, townships and work units were transferred, by payment, to a state-run city-building business. The absence of democratic politics meant that there was little to distinguish this new-city building project from the vast numbers of smaller private neighbourhood-building projects currently shaping the commodified outer rings of Chinese cities.

Beyond the urban fringes, the redistribution of state land had a significantly different outcome to that within the city. Villagers living within commuting range of a large city, like many city dwellers, suddenly became potentially wealthy overnight. The rights that were transferred to villagers were not alienable on the open market, however; land was collectively owned and could not be legally sold on while it remained legally registered as rural. Village land, including agricultural land and the land on which village housing is built, can, however, be converted to urban use if an application is granted by the municipal authority into whose jurisdiction the village would accede. To realise the market value of their jointly owned suburban and ex-urban land, villagers must sell their birthright to the municipality. This is a conundrum that has produced great inventiveness. The stakes are high and all parties stand to benefit from negotiation. Municipalities own the right to pronounce land within their territory 'urban', villagers collectively own land-use rights within village territories, developers own the know-how to create saleable urban land and buildings, and investors own capital. We demonstrate later some of the outcomes of this interesting game. In most if not all cases, the result is a reconfiguration of the co-owned realm, demonstrating the surprising resilience of condominium as an institutional form in cities.

One way in which villages or rural townships found they could capitalise on their latent land-based wealth was to set up a township and village enterprise (TVE). This is a corporation in which villagers or farmers have an ownership share and which operates some form of enterprise. The nature of the enterprise varies by location and natural resource endowment. Many engaged in extractive industries, including coal. Villages lying in the shadow of Shenzhen and Hong Kong developed speculative factories and worker dormitories, which they rented on short leases to domestic and foreign manufacturing investors – even for one-off production batches. Villagers reaped the benefits of their land and location through factory and dormitory rents and from income from associated services and commerce. The result was intensive and discontinuous development of village land and highly unusual spatial patterns of economic activity. Villagers have the incentive to maximise investment and use of land within the communal area since they cannot legally transfer their land-based wealth outside (by selling it) and if they move out of the village they forfeit their part-ownership of the collective enterprise.

On the boundary of the city, particularly in Beijing, lie new forms of condominium: China's golden enclaves. Guillaume Giroir's chapter in this volume paints a vivid description of one of the earliest examples. Virtually all high-end housing in China is gated and organised under private managerial governance.

The wealthy gated communities occupying favourable locations in the north-east of Beijing are little different to those found in cities throughout the world. They are particularly novel in the Chinese context because income differences have only really been conspicuously revealed in spatial patterns since the housing market was created in the 1990s. They are, perhaps, novel in an international comparison of high-income gated developments because of the scale of the otherwise familiar design features (figure 11.1). Other than this they are much like any other imported global consumer product (Wu 2004).

What is of more interest is not their difference compared to other forms of commodified housing in Chinese cities – largely a matter of design and location – but what they have in common. One of the most startling features of post-reform urban China is the widespread adoption of private and partnership neighbourhood governance. While one would expect this of the *golden ghettos*, the widespread devolution of neighbourhood management to entrepreneurial organisations is unexpected to the outsider. To the inhabitants of China's cities, private or semi-private neighbourhood governance – and the walls and gates that go with it – is unremarkable, however. People have long lived in neighbourhoods organised by agencies other than the municipal or district government (Wu 2005) and the growth of a competitive market of proprietary neighbourhoods is a natural accompaniment to the commoditisation of housing.

Work unit residential neighbourhoods were typically built and managed for *dan wei* by local governments or other state agencies under contract. If they were not managed by a local government agency they would be managed by an enterprise's own estates division. When property rights began to be transferred to tenants after 1997 this started to change. Collective action typically became organised through property owners' associations that contracted governance and managerial functions to entrepreneurial management firms (Pei 1998). The latter include former estate management units of state enterprises and local governments. Most estate-based housing is therefore contractually rather than politically governed. This renders the distinction between high-end gated communities and median-income housing estates less fundamental than in many other countries. They are all organised as various forms of condominium and in this sense the Chinese urban governance experiment seems to be rather unique – although there may be some parallels in the transformation of housing cooperatives in central and eastern Europe.

This is most strikingly illustrated by large suburban master-planned mixed-income communities. Later in the chapter we describe one of these in the central Chinese city of Wuhan. The size of a small city in Europe, this is privately funded, walled and gated, and organised under contractual condominium institutions. In addition to the commoditisation of housing and physical neighbourhood infrastructure the commoditisation of community services – including the supply, management and governance of services – is widespread in contemporary urban China.

Figure 11.1 Golden ghetto on the outskirts of Beijing. Mostly retired parents of successful young professionals live in this development, which offers a hospital, among other facilities, behind its neo-classical gateway, 2004. Photo: Chris Webster

Snapshots of China's diverse condominia

A work unit neighbourhood

The estate financed by the education bureau of one of Beijing's districts in the south-west of the city was one of thousands of similar developments constructed in the city in the 1980s. The bureau built around 1,000 units in a cluster of walled and gated estates of about 200 each (see figure 11.2 overleaf) and offered these to high-performing teachers from nearby schools, most of who lived in poor-quality dormitory housing on school sites. The bureau's own estate management division supplied neighbourhood services, including a resident concierge who acted as general neighbourhood manager, residents' representative, education bureau representative, guard and gatekeeper. The local streets between the walled estates were maintained by the local street office. Twenty years later, the teachers, walls, gates and gatekeepers are still there, only the latter now serve an additional function: private letting agents. Beijing was late in implementing the 1997 housing reforms and the teachers resident on this estate started to purchase the use rights of their apartments in 2001. When the authors visited one of the estates in March 2004, most residents had paid the bureau for their apartments but were still waiting for deeds. The gatekeeper, himself now an owner–occupier with a property owning interest in the quality and status of renters, screens the daily stream of mostly migrant workers seeking somewhere to rent.

Since teachers were allocated their apartments on the basis of performance

Figure 11.2 Brown ghetto in inner Beijing. One of Beijing's many drab five-floor walk-up gated compounds built by work units in the 1980s. This one is home to a thousand retired teachers – a Chinese-style common-interest development, 2004. Photo: Chris Webster

and rank, most were senior then and are now retired. The estate is a peculiarly Chinese version of North America's retirement common-interest developments (CIDs): a walled community of 1,000 retired teachers and their spouses and, in some cases, children – now students or young professionals. Asked why it is that very few have moved away, five non-mutually exclusive explanations emerged: first, the strong sense of community; second, the advantageous location – within the third ring road and near bus and rail stations; third, the delay in receiving property deeds; fourth, the strategic decision to wait for a grown-up child to make a career or home-purchase decision before deciding whether to keep or sell; fifth, the rising investment value of the properties. Interviewed residents were clearly aware of the advantages of staying put while property values rose by double-digit figures. Most residents purchased their 50 to 70-square-metre apartments at the highly subsidised price of about 20,000 yuan (US$2,300). Three years later they were typically worth around 300,000 yuan. Asked what would make them move, most quickly offered the problems of using stairs – these are the five-storey walk-up blocks ubiquitous throughout China. This highlights an urban management time bomb. The many 1980s-style estates like this 'teachers' quarter' are of very basic design and do not compare with the city's commodified housing. They are ripe for redevelopment, renovation or gentrification, or for decaying into slums. The last option is less likely than in many cities because of the unified ownership of land and common facilities.

This particular condominium story is reasonably typical of the 1980s ring of rationalised work unit housing and the comprehensively developed suburban housing districts built in the same period by partnerships of work units. Some of these estates have been more responsive to the rental market – some having clearly passed a tipping point where the proportion of 'unknown' renters leads to an accelerated exodus of original residents. They share a common collective ownership heritage, however, that is translating into the market era in interesting ways. The collective assets and common services in the teachers' neighbourhood are still owned by the district education bureau and managed by the bureau's estates division. Residents currently pay no service fee and the arrangement is a hybrid between municipal housing, corporate-owned housing and modern condominium housing. It is not local government housing and it is not entirely private – only the usufruct rights to individual units are in private ownership. Other estates, such as the Fangzhuang estate in the Chaoyang district of Beijing, originally built for 30,000 employees of the Water Resource Ministry and the Ministry of Foreign Affairs, have contracted with private management companies to supply local public goods and services, making them more like the modern contractually managed neighbourhoods. One suspects that the homogeneity of the teachers' neighbourhood and the high level of loyalty (both ways) between residents and former employer has prevented a more contractual condominium institution and organisational arrangement from emerging in that estate. When the teachers start letting or selling, this is likely to break down and formal contracts are likely to replace informal conventions based on loyalty and trust. The relatively stable community structure is due to perhaps two reasons. First is the legacy of welfare-type development. The project started as *dan wei* welfare housing provision and the education bureau is still viewed as a paternalistic organisation looking after its employees, notwithstanding the various market elements introduced into the management. Second is the relatively low level of capitalisation seen in this neighbourhood. Not only are residents still waiting for deeds but also the potential capitalised value is far below the 'decent' commodity housing that matches the teachers' aspirations. The average level of 600,000 to 700,000 yuan for well-serviced commodity housing at an equivalent location is clearly not affordable for this group who are better off than the average but still face a 300,000 to 400,000 yuan price gap. The age of residents means that a mortgage is not an option for financing this shortfall. In view of the constraint of consumption capacity, neighbourhood membership based on affiliation rather than an explicit contractual relationship may be more efficient, as members clearly cannot afford fully commoditised community services.

'Village' condominia

Three examples illustrate the evolution of collective village rights under market-driven urbanisation.

Village 1 Outside the Special Economic Zone of Shenzhen, but within the territory of the new city, is the 'rural' district of Baoan. Within Baoan there are eight towns and 113 villages and about 80 per cent of land in Baoan is owned by the village collectives. Shajing town in Baoan District contains 26 villages within its territory, organised as a single villager committee with over 26,000 members. By the end of 1997 approximately 4 million square metres of industrial workshop space and 1.5 million square metres of worker dormitories had been constructed, accommodating more than 800 firms with an annual output in 1997 of approximately 3.2 billion yuan (US$364 million). While villagers are constitutionally constrained from selling their land rights and cannot legally convert land to urban use without losing ownership, they can get round this by giving their rights to a village enterprise corporation. In the case of Baoan, factory and dormitory space was constructed by the collective enterprise and rented largely to Hong Kong manufacturers. Only 17 per cent of industrial output in 1996 came from local industry. By 1997 in one of Shajing's villages, Bogang, 1,550 villagers were renting 150 workshops and 163 dormitories to over 100 firms and making over 90 per cent of village income from this activity. Villagers gained further income from supplying local infrastructure, goods and services – which also functioned as an investment that made the village more attractive to mobile industry. Between 1991 and 1995 the village invested more than 30 million yuan (US$3.5 million) in infrastructure such as roads, water and power. In 1996 it invested 7 million yuan on new roads, 2 million yuan on a park, 3 million yuan on a clinic and 4 million yuan on a school. It also supplied a private security force of 60 and a refuse collection force of 50.

Because of China's dual land-use system, in urbanised 'rural' areas such as this, the local village or township governments actually act as the municipal government and become the 'landowner' of rural land. Legally, the lease of rural land is controlled by the counties but the system allows villages to convert their rural land into 'non-agricultural uses' via TVEs. Land used for farmers' housing (*zai ji di*) is strictly speaking constrained and cannot be sold in the urban market, and neither can the 'non-agricultural land' be transacted in the urban land market. In reality, however, the duality has created a black market of rural land. The black market, like the TVEs (and pre-land-conversion negotiation between village committees and developers – see Village 3 case study below), subverts what is otherwise a municipal monopoly right over land betterment value – achieved by the land-leasing system. Municipalities have tolerated these leakages of development betterment to rural farmers but competition for control over land uses in urban villages has become a seriously contested issue.

Village 2 Forty kilometres south-west of the Fangshan district of Beijing is Hancunhe, a village of 910 families and 2,712 people (2002 figures, collected by the authors during a field visit in 2004). In 1984 the villagers, who had organised themselves into a construction workforce that started with 30 workers and quickly grew to 1,000, broke into the lucrative Beijing market by winning a contract to construct a 8,400-square-metre hotel. Dozens of major projects later, the village construction enterprise now employs something like 50,000 workers in the city and Hancunhe has transformed itself into a model high-density ex-urban village. The village enterprise production value was 2 billion yuan in 2003 (over US$200 million). Villagers now live in 581 villas (see figure 11.3 overleaf) and 21 well-appointed apartment blocks – a half-billion-yuan investment by the village collective on behalf of its members. The new homes were allocated through a subsidised internal-village market arrangement. Families were each allocated one square metre of new accommodation for every square metre of old building occupied (all of which were demolished) plus a 30,000-yuan subsidy. Families added their own savings or borrowings to this to bid for different categories of accommodation. All of the once-poor farmers have a television, refrigerator, washing machine and video recorder and more than 80 per cent of families have computers.

The village also invested in collective civic goods, including a 30-million-yuan (US$3.5 million) education centre housing kindergarten, primary and senior schools and a junior college. It continues its rural collective welfare tradition by, for example, making monthly or annual payments to the over 60s (2 yuan per month); primary-school-age children (100 yuan per year), the disabled (720 yuan per year); and parents of only one child. The village subsidises the collective heating service to the tune of over 5 million yuan per year, while village members contribute approximately 185 yuan per year per family.

Construction enterprise income has been used to diversify the village's economic base as well as improve the standard of collective and private consumption. In 1996 it invested 25 million yuan in a 6.5-acre high-tech market garden producing quality and exotic products for the Beijing market. A travel agency, a hotel, a bowling alley, a gym, tennis courts and a luxury swimming pool have consolidated the tourist business that seems to have grown partly on account of the village's 'rags-to-riches' story. The village is also accessible to a famous cave, temple and mountain landscape areas and now markets itself as 'Hancunhe tourism scenic village' (see http://www.hancunhe.com.cn).

The latest collective enterprise is a natural extension of the hotel and tourism business – a 3-million-square-metre high-rise condominium development on the edge of the existing village area. By 2004, 2,300 of the 5,500 condominium units had been completed, with 80 per cent of buyers being Beijing residents purchasing a second home in the country (for between 138,000 and 308,000 yuan – US$15,000 to 30,000). The commodity housing development will have involved the reclassification of this part of the village from rural to urban land and the transfer of rights from the village collective to the municipality. Villagers have managed to capture rent from the lost asset, however, through their ownership of the construction company building the apartments, through the local trade generated by the new occupants and through ownership of the property management company. In the long run, the village stands to gain from this early agglomeration of urban land uses; the

Figure 11.3 Green ghetto in the countryside 50 kilometres from Beijing. In this village 'condominium', farmers-turned-construction-workers redeveloped their collectively owned land, building each family a villa and diversifying their village enterprise by constructing high-rise apartments for sale to non-villagers, 2004. Photo: Chris Webster

county, which is within the Beijing municipal territory, has recently been upgraded to an urban district and it will not be long, we predict, before Hancunhe becomes an outlying sub-centre in the greater Beijing metropolis.

Hancunhe is unusual, if not unique, in terms of its economic diversification into property development. A powerful and capable local elite, building upon their political capital, might be an explanation why this village could so successfully tap into Beijing's construction market. The ability to redistribute such large benefits so conspicuously to the village 'citizenship' derives in part from the initial political capital endowment, the successful TVE economy, the teamwork and leadership that produced the early gains, and the strong ongoing commitment to collective ideals. This is in sharp contrast to most TVEs in southern China where benefits are either transferred into private corporations or diminished by tough trading conditions.

Village 3 In the shadow of Beijing's walled Summer Palace sits a village of 4,000 to 5,000 people. Half of these (about 1,000 households) are the original erstwhile farmer–villagers and the other half are a floating population renting from ex-farmers, retired PLA soldiers who have been allocated houses by the government, and other more recent incomers. Until January 2004 the process of converting rural to urban land in Beijing permitted village committees to negotiate deals with developers before applying to the municipality for change of use (and thereby relinquishing ownership). The rural residents sharing the collective land rights to the Summer Palace Village set up a company that entered into a joint venture partnership with a developer to construct the Shanshui Yijing project. The village is in a prime location and competition between developers meant that favourable terms were secured by the village committee. This included entitlement to a compensation package of 40 to 50 square metres per person at 5,000 yuan per square metre – either taken in cash or in a part or wholly in-kind subsidy in the form of alternative accommodation on an estate built by the developer less than two kilometres away. Villagers wanting to upgrade beyond the typical package, worth 400,000 yuan per couple (US$45,000), could make additional payments. This package was available to villagers with shared usufruct rights (excluding renters). The collective's original farming households had the additional entitlement of a joint (and inseparable) share in the joint venture enterprise – which included a property management company that would service the new villas built on the site. As a result of public controversy over developing such a sensitive site and the municipality's policy of extending Beijing's green belt in the run-up to the 2008 Olympics, the municipality insisted on four to five acres of the converted land being turned over to forestry.

The farmers, acting as monopolistic collective landowners, levered a significant individual pay-off package, the right to maintain the new forested park, a share in the development profits via the joint venture company and a share in the business of managing the new estate of 500 or so up-market villas and apartments (selling at 5 to 6 million yuan per unit). Approximately 80 per cent of the villagers chose to relocate together in the alternative accommodation constructed by the developers, electing to maintain the condominium institution that had served their interests so well. The developer of this project noted that most farmers strongly support the idea of co-ownership because the collective gives them more benefits in the form of enhanced social security and a strong voice in negotiating windfall dividends from land inherited from the socialist era. The villagers' favourable deal from the project is partly due to the specific context in which the Beijing municipal government seems to be less able to monopolise the primary land market (a situation in part due to the location of powerful state interests in the capital). The good deal may also be explained by the pressing need to develop a green belt in time for the Olympic Games and greater willingness to give property developers flexibility in land acquisition.

Master-planned private suburbs

The industrial city of Wuhan in the central Chinese province of Hubei provides one of the most noteworthy examples of modern contractual government

Figure 11.4 Red ghetto in the suburbs of Wuhan. This is the central square in a private town, built for over 200,000 by a private for-profit company who runs it under a strong communist ethos, 2002. Photo: Chris Webster

in the world (Liu 2003). In the early 1990s, the Wuhan city plan allocated 1,000 acres of partially derelict agricultural land on the outskirts of Jiang District for new housing. After various withdrawn bids from state-owned enterprises the Wuhan municipal government signed a contract with the Wuhan Housing Project Development Corporation Ltd (WHPDC), a wholly owned subsidiary of the privately owned Anju Corporation. Anju's owner and chairman, Mr Mao, had made his fortune in the early days of south China's property boom and was keen to invest in Wuhan. The principal contribution of the municipal government was to organise the re-registration of the land from rural to urban, to negotiate villager relocation and compensation and to transfer leasehold rights to WHPDC/Anju. Thereafter, the design, construction, management and governance of Baibuting Residential Quarter was the responsibility of a for-profit company. The initial investment included 300 million yuan for road, pipes and other infrastructure to support a planned target of 150,000 to 200,000 people. This included a 70-million-yuan, 30,000-square-metre green square at the centre of the town (figure 11.4), three kindergartens, two primary schools, two high schools, a clinic, sports facilities and an administrative centre. By 2003, 60,000 people had moved in; a private urban administration of 140 was employed by the private town hall (office of the property management company and residents' committee), including 20 managers/administrators, 30 maintenance staff and 60 security officers; and work was under way to develop the remainder of the 3.5-square-kilometre building area over a seven-year period. The total scheduled investment over a seven-year period is US$200 million, with roughly a third coming respectively from Anju's equity, banks and property sales. The project broke even after five years and turned a US$2 million profit in 2001, approximately 5 per cent of which was placed in a reinvestment fund. Home-buyers are median-income residents – prices per square metre were 908 yuan in 1995, rising to about 1,500 yuan per square metre in 2002 (around US$8,500 for a 50-square-metre apartment, for example). Residents have a say in the government of the town, including allocation of the monthly service fee income and the reinvestment fund, through the residents' committee. This is combined in a 'three-in-one' community management model – a unified community service centre, property company and residents' committee. The relationship with municipal and district governments is a reversal of the more usual partnership model in other countries: Anju contracts with the best schools and hospitals in Wuhan to supply educational and health services to its residents in its own buildings. Similarly it provides accommodation to six municipal police officers who partner the private security force. The town organises its own welfare service, including free health checks for the over-60s and under-12s, scholarships and employment opportunities for laid-off workers.

The entrance to Baibuting is not gated – there are publicly accessed banks, supermarkets, shops and informal market stalls along the town's main private arterial road – but all of the town's housing areas leading from the main roads are separately walled, gated and guarded.

Baibuting is a good example of market-based proprietary neighbourhood development that succeeds by tying in an obviously attractive bundle of community goods with private housing. Residents interviewed during a 2002 visit by two of the authors described their search for commodity housing and the importance of the amenity package. It was clear in talking to residents and property managers that the expectations of the town were shaped by the shared experience with *dan wei*. Indeed, the loyalty expressed was not dissimilar to that found in the retired teachers' estate in Beijing. A palpable sense of local social contract was discernible in many of the residents' comments – including a trust in the town's management to act in the residents' interest. This expectation had so far been met, with several examples of significant investments made in response to residents' demands (for example turning a piece of open space into a basketball court). The redistributive services of the town were a matter of pride and a selling point for the management. Welfare services and redistribution within a social unit less than the municipality are again reminiscent of the paternalistic ethos of the work units. Such an example is not unique in China, as the government actively seeks to use commercialised community services to subsidise non-profit activities. In Baibuting the scale of the community is sufficiently large to achieve economies of scale and to distribute the costs of welfare sufficiently thinly. This, plus unified ownership and governance, permits the town to plan and cost welfare into its product specification and marketing strategy. The usual problem for mixed social and profit agendas is that the scale is too small to achieve economies or to seal the boundary against leakage.

Conclusion

Baibuting is not the largest privately run gated town in Wuhan – another, Chunchin, seen from the privately supplied airport toll road, has a planned target population of 300,000. Gates are unremarkable in China. As one developer put it, 'walls are the tradition here'. This was so historically but became truly institutionalised during the central economic planning era. Surprisingly to the outside observer, walls, gates and the condominium governance that typically accompanies them has emerged, or more precisely been 're-discovered', as the dominant form of order in China's new wave of commodified housing development. China and the USA together lead the world in terms of the scale and pace of private urban governance innovation. China comes from a very different set of starting conditions, however. In the USA, contractual government and private planning are emerging as the dominant form of new development. This is so in China too, but there most older housing in cities is also under various forms of condominium rule. The socialist planning era had the effect of embedding a tier of urban governance that failed to emerge strongly if at all in 20th-century Western cities. Contrary to the commonly perceived view that the socialist city was under strict centralised public ownership, socialist management was much more ad hoc and operated through a wide range of cellular structures with varying elements of communal ownership. Economic reforms injected more locally based resource control into these structures and transformed a territorially fragmented autocracy into a competitive market of micro-spatial governments. The emergence of China's modern 'gated' urban ecology is explained by the infrastructural and institutional legacy of socialism which delineated the boundaries of joint consumption rights far more clearly than was the case in non-socialist cities. What the market is now doing is delineating private consumption rights in a way that socialism could not. The result is a city of residential *club realms* governed by varying models of condominium.

We speculate that there is, in the two super-economies of China and the USA, a similar, possibly converging, evolutionary path of intuitional, organisational and spatial urban order. Neighbourhood-level civic goods and services are plainly effectively supplied to 'small publics' – groups of co-consumers greater than the individual or household and less than the whole population of an urban jurisdiction. The contemporary Chinese walled *cities within a city* are all the more interesting in that they are a genre of development adapted to all income levels – from the poor still living in courtyard housing to the rich in Beijing and Shanghai's Californian-style residential theme-parks, and the old- and new-style middle-income condo dwellers in between. The mix of co-ownership and selective territorial enclosure is a powerful formula with an apparently compelling economic rationale. Research is needed to understand how this economic rationale shapes space and lives in different political economic realities.

Bibliography

Buchanan, J. (1965) 'An economic theory of clubs', *Economica*, 32: 1–14.

Cornes, R. and Sandler, T. (1996) *The Theory of Externalities, Public Goods and Club Goods*, 2nd edn, Cambridge: Cambridge University Press.

Friedman, J. and Weaver, C. (1979) *Territory and Function: The Evolution of Regional Planning*, London: Edward Arnold.

Heikkila, E.J. (1996) 'Are municipalities Tieboutian clubs?', *Regional Science and Urban Economics*, 26: 203–26.

Liu, J. (2003) 'Wuhan's Bai Bu Ting Residential Quarter', in C. Webster (ed.) *Public–private Partnerships in Urban Infrastructure and Service Delivery*, Bangkok: UNESCAP: 57–9.

Pei, M. (1998) 'Chinese civic associations: An empirical analysis', *Modern China*, 24 (3): 285–318.

Webster, C. (2002) 'Property rights and the public realm: gates, green belts and Gemeinschaft', *Environment and Planning B*, 29 (3): 397–412.

Webster, C. and Lai, L.W.C. (2003) *Property Rights, Planning and Markets: Managing Spontaneous Cities*, Cheltenham, UK and Northampton, MA: Edward Elgar.

Wu, F. (2002) 'China's changing urban governance in the transition towards a more market-oriented economy', *Urban Studies*, 39 (7): 1071–93.

—— (2004) 'Transplanting cityscapes: The use of imagined globalisation in housing commodification in Beijing', *Area*, 36 (3): 227–234.

—— (2005) 'Rediscovering the "gate" under market transition: From work-unit compounds to commodity housing enclaves', *Housing Studies* 20 (2): 235–54.

Zhang, T. (2002) 'Decentralization, localization, and the emergence of a quasi-participatory decision-making structure in urban development in Shanghai', *International Planning Studies*, 7 (4): 303–23.

12

THE RISE OF GATED RESIDENTIAL NEIGHBOURHOODS IN PORTUGAL AND SPAIN

Lisbon and Madrid

Rainer Wehrhahn and Rita Raposo

Abstract

Gated and pseudo-gated residential neighbourhoods took root in the 1970s in Spain and in the 1980s in Portugal. Analysing these two countries is not only interesting because they are among the first European countries admitting private housing estates, but also because they underwent a relatively late post-modernisation process within their urban realms. The main objectives of this paper are to present a first empirically based analysis of the structures as well as of the reasons for the rise of both gated and pseudo-gated communities on the Iberian peninsula, using the metropolitan areas of Lisbon and Madrid as examples.

Although some similarities between both countries have been found, especially when linking the phenomenon to the process of suburbanisation and to post-suburban processes, we revealed significant differences with respect to explanations of the gated-community phenomenon in the context of societal and institutional change and spatial developments. In Madrid, for instance, in contrast to Lisbon, the first larger gated communities were conceived as weekend resorts for the better off during the 1960s and 1970s and then were transformed into permanent resident settlements in the last two decades. Links between political and economic events or periods can be found in Madrid. With respect to the weak-state hypothesis, we found some evidence in both case studies, but only in Portugal does the search for amenities the public sector cannot provide – as for example in the US and in Latin American cases – seem to be one explanatory factor for the proliferation of gated communities. In both Lisbon and Madrid we revealed another important reason: social distinction via walls, private security employees and prestigious club goods must be interpreted as a decisive factor for strictly gated as well as for only symbolically gated communities. But in contrast to North American private neighbourhoods, the Iberian cases showed very little evidence of the wish for private governance and for the 'community' aspect as explanations for the rise of gated communities.

Introduction

Gated private neighbourhoods first took root on the Iberian peninsula in the late 1970s and early 1980s, with some earlier Spanish exceptions dating to the 1950s. In Spain and Portugal this specific socio-spatial form, whose origins can be traced back to the beginning of modernity, seems to be merely a late 20th-century novelty and an import, even if one whose paths are not particularly easy to determine. Gated neighbourhoods, in their modern form, began within the Anglo-American world, resurging in the United States in great numbers in or around the late 1960s. They have now spread worldwide proving to have, in many countries and regions, a range of affinities with what has been termed postmodern society, space and urbanism. Gated communities, in their earlier forms, arose in a transitional, recent modern, bourgeois and liberal capitalist Anglo-American social-spatial setting. Two different but sequentially and socially connected residential forms appear to make up this first version: the enclosed and gated version of the mid-18th century British residential square, and the planned mid-19th-century Anglo-American *romantic suburb* (Raposo 2002, 2003). Later instances of this phenomenon or its contemporary form are much more extensive and can be found in a greater and more diverse world, including Spain and Portugal.

Probably from the beginning, and surely from the romantic suburb form onwards, gated housing communities (GCs) have simultaneously been a unique form of segregation and a commodity, a real-estate product (Raposo 2002, 2003) and an established commercial formula. As a real-estate product, GCs are the direct outcome of the real-estate market structure, its dynamics and agency. They are heavily subject to the aestheticisation logic that dominates, especially in contemporary society, the production, circulation and consumption of economic goods (Lash and Urry 1994). They further represent a good example of the progressive commodification of social life, namely in the spheres of habitat, citizenship and local government. As a form of segregation they can be distinguished from other known forms of segregation through the unique association of two traits: their use of physical–architectural barriers and their voluntary character. This unique form, which represented a very limited variety of modern segregation (in extension and duration), is today much more important but it also reflects profound transformations in *space* and *society*. It can be specifically associated with a number of diverse countries with experience of economic, social and spatial restructuring and of postmodern urbanism, culture, society and space (Davis 1990; Soja 2000). More generally, it can also be related to social polarisation and insecurity, but also to a 'weak state', i.e. to the failure or the abstention of state intervention in the regulation of space and society and in the provision of different kinds of public goods.

Analysing Portugal and Spain is not only interesting because both countries underwent a relatively late (post)modernisation process within the western European context but also because our case studies reveal that despite certain obvious similarities concerning political and societal changes in the 1970s and 1980s, the explanation of the rise of gated communities in these southern European countries seems to be much more complex than it would at first glance appear. The transition from dictatorship to democracy since 1974 and the integration of both countries into the European Union 12 years later led to similar but not to an equal gated-community panorama. As a consequence of this finding, documented earlier by both authors (Raposo 2003; Wehrhahn 2003), our main interest now is to compare in more detail supply and demand in the context of globalisation processes as well as in connection with the national and local set of historical, political, social and economic changes experienced since the 1970s.

Our analysis focuses on gated residential neighbourhoods which are defined as a residential form that may cover differing housing forms – detached or semi-detached houses, rows of houses or apartment blocs (or any mix of such forms). The definition incorporates the following characteristics:

1 private or privatised amenities for collective use (streets, green areas, swimming pools, tennis courts, golf courses etc.);
2 perimeter impermeability and access control (Luymes 1997);
3 collective private ownership of open space or privatised access to it.

The alternative we created around the categories 'private' and 'privatised', relating to 'amenities' and 'open space', is due to the fact that there are in the Lisbon Metropolitan Area as well as in Greater Madrid gated housing estates that were built as walled and gated estates while retaining within them public spaces such as streets and green areas (not coinciding with private housing or equipment lots). Those developments are legally open to anyone minded to traverse them, although in practice only a few people know they hold the right to enter these privatised spaces. This peculiar situation, which corresponds to what we call 'pseudo-gated communities' (Wehrhahn 2001: 108), has thus far not been reported in any other case apart from in Madrid and Lisbon (Wehrhahn 2001, 2003; Raposo 2002, 2003).

Methodology

For the study of gated housing estates in the Lisbon Metropolitan Area (LMA), an area-wide survey of the phenomenon was carried out covering the period from 1985 to 1999. This included a development census involving the identification of existing cases and the construction and completion of a database covering fields of several types. Consultation and analysis of the specialised press and specific legal texts, observation of several developments and in-depth interviews with key participants such as real-estate agents, municipal council technical personnel, architects and legal experts, completed the main methodological procedures that were used (for more details see Raposo 2002). For the metropolitan area of Madrid standardised interviews involved inhabitants of three different types of gated community: strictly gated communities, pseudo-gated communities and smaller units called condominiums in this case study (see below). The interviews served as basic information on the demand side, i.e. the motivations of families migrating to gated communities. A second information source has been expert interviews with representatives of planning institutions and gated-community administrations, with security personnel and real-estate salesmen and representatives from 1999 to 2001 (Wehrhahn 2001). Since there has been no published research about gated communities in Spain this study has to be regarded as a sort of pilot study, emphasising the generation of new hypotheses rather than proving those existing.

Profile of gated communities in Lisbon

The Lisbon Metropolitan Area has 19 municipalities over its 3,122 square kilometres, with approximately 2.5 million inhabitants, and a very unbalanced settlement pattern in municipal terms. Throughout the aforementioned period, a total of 97 gated housing estates in the LMA were identified distributed across just ten of the municipalities (see figure 12.1 overleaf). Furthermore, since the beginning of 2000 and at least through to the end of 2001, the phenomenon was clearly spreading in popularity and to such an extent it reaches, even if at a distance, areas of other LMA municipalities. Those two years corresponded to the second phase of a positive cycle for the Portuguese real-estate market. Starting in 1998, the property boom coincides with the point at which *condomínios fechados* (GCs) became quite a common and even fashionable real-estate product in Portugal. From 1998 onwards an important and growing part of the new housing stock targeting middle and upper middle segments of the market has been gated. Furthermore, if the metropolitan areas of Lisbon and Oporto, and the tourist region of the Algarve, have for a long time been the almost exclusive locations of gated housing estates in Portugal, they may now also be found at differing geographical locations across the country, including small towns and different kinds of tourist destination. The scale it has reached, its proven commercial success and the social visibility attained seem to have encouraged its expansion, including through the targeting of lower real-estate market segments. Finally, the translation of the concept from the expected market of luxury or high-to-medium standard housing to the lower-priced segments has its 'natural' limits, particularly when it comes to an economic turndown. By 2002 another negative cycle for the Portuguese real-estate sector in particular and for the economy in general had begun. This still ongoing conjuncture is limiting the expansion of housing production, including that of GCs, which also proved in previous periods to be at least partially sensitive to economic cycles.

Of the 97 developments, only 10 were dated prior to 1990, and all were

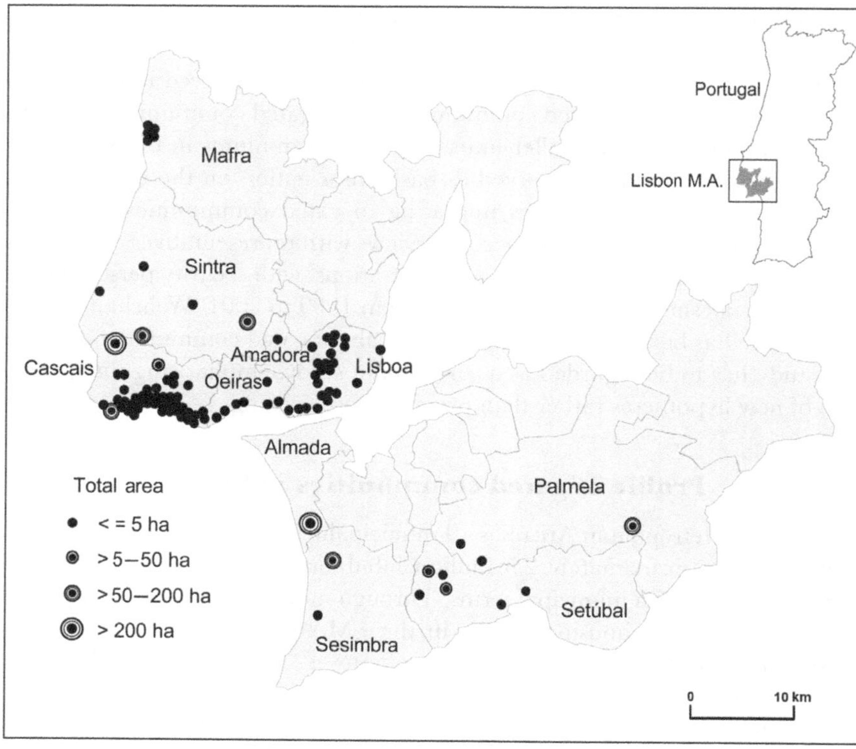

Figure 12.1 Location of gated communities in Lisbon Metropolitan Area. Source: R. Raposo and F. Honorio

small-scale developments with areas of less than five hectares. In the following three years, a further 27 new developments were built in the LMA, including the first large-scale and master-planned gated housing estate. Over the next five years the real-estate market experienced a sharp slowdown with the growth rate of this residential form also declining. Indeed, there were only 28 new gated housing estates built during those years. Finally, in the last two years under survey, the years 1998 and 1999, the real-estate market recovered, helped by a very positive macro-economic conjuncture and by significant and steady decreases in mortgage interest rates. During those two years a further 32 new gated developments were built in the LMA.

Concerning the surface and housing units making up those developments, 43 out of 97 developments occupy a total surface of 1,424 hectares (as for the other 54 cases, we know only that each project is smaller than five hectares), and 79 cases account for a total of 10,571 housing units. The surveyed universe of 97 gated housing estates presents highly contrasting formal characteristics. Besides the fact that only 10 are over five hectares in area, what is noticeable is the great differences the universe presents in terms of the number of housing units, the types of amenity and the type of security used. In fact, if we compare only

Figure 12.2 The access gate of a master-planned gated housing estate located in the municipality of Cascais. Photo: Rita Raposo

the cases of these small-size developments we can observe that in that universe there are developments with less than 20 housing units and some that reach almost 300 units. In general, only the big (ranging from 100 to 354 hectares and from 200 to 1,633 housing units) and medium (ranging from 16 to 46 hectares and from 30 to 243 housing units) developments have the scale required to support space demanding expensive amenities, like golf courses, parks, tennis courts and schools. In the case of the smaller developments, the more common amenities are swimming pools and gardens. The 'heavier' security systems are typical of the developments with a larger number of housing units.

Through multiple-correspondence statistical analysis, it was possible to identify the existence of three distinct classes of gated development in the LMA. Briefly, the first class (72 cases) is characteristically a class composed of apartment complexes (of one or more buildings) of between 21 and 50 apartments or between 101 and 200 apartments. The surface is equal to or less than five hectares and they lack any amenities like golf courses, tennis courts or clubhouses. In the Lisbon municipal areas there are only developments of this type. The second class (18 cases) is characteristically a class composed of developments with detached or semi-detached houses and less than 50 units. Finally, the third class includes seven cases that correspond to the totality of major master-planned gated housing estates in the LMA (figure 12.2). More characteristically, the developments of this class are distinct given their range of amenities (all those that negatively characterise the first and second classes in addition to others), their number of housing units, their area and their mixed housing character. The second and third classes are typically found in

Figure 12.3 A small gated sub-development inside a pseudo-gated master-planned gated housing estate located in the municipality of Sintra. Photo: Rita Raposo

suburban municipalities, even while these also display developments of the first class. In all three 'morphological' classes a few cases of pseudo-gated communities are to be found but the great majority are strictly gated. Only among the third class is that variety significant – three in seven developments are pseudo-gated. Moreover, two of these communities are home to sub-developments, usually small apartment complexes that are strictly gated (figure 12.3).

Looking further at the spatial organisation and settlement patterns of the territory of gated housing estates, one can observe that small-scale developments are located in 'areas of dense and continuous occupation' (Salgueiro 2001: 118) and that the big and medium-scale developments are located outside the urbanised perimeter of the suburban localities, in areas that may be described as 'areas of intercalate urbanisation' (ibid.). Furthermore, it is worth stressing that the municipalities of Cascais, Oeiras and Lisbon, forming a continuum along the northern bank of the river Tagus, are home to, respectively, 43, 7 and 23 developments. These alone amount to 73 cases. The Lisbon–Oeiras–Cascais axis, usually referred to as the *Linha de Cascais*, corresponds to the most affluent and most socially favoured LMA area, according to several socio-economic indicators and studies (INE 1999; Salgueiro 1997). This axis has a traditional prestige related to both its landscape qualities and its image of affluence, cosmopolitanism and social distinctiveness. Following the same logic, inside Lisbon, the *condomínios fechados* are located either in old, consolidated and prestigious areas of the city, or in its new or expanding areas where social images are still under construction with no 'negative' symbolical connotations such as working-class, mass-market or decaying and derelict.

Clearly, gated housing estates both across the LMA and inside Lisbon are preferentially located in more affluent, expensive and prestigious sites consistent with their targeting of the middle and upper middle classes. In Portugal, and concerning first homes, inhabitants of gated housing estates are typically middle and upper middle class (members of the upper class, having greater freedom of choice, seem to generally prefer more 'individualistic' housing solutions). As such, they seem to enforce pre-existing segregation patterns inside Lisbon and, in more general terms (as the scale changes), in its metropolitan region. However, it should be noted that Lisbon, like other Mediterranean metropolises, has comparatively lower levels of socio-ethnic segregation and functional differentiation than northern European metropolises (Salgueiro 1997) 'due to the late development of industrial capitalism, and of a formal urban planning culture' (Malheiros 2002: 115). At the same time, 'the relative exceptionalism of urban regulations' results in the 'cities of the South and their suburban extensions displaying more "disorganised" patterns, in both urban and social terms' (ibid.). As such, gated housing estates seem to provide an additional barrier against a territory that may be experienced as 'insufficiently' socially clustered and separated.

Explaining Lisbon's gated communities

To explain the rise of private gated residential neighbourhoods in Portugal, and particularly in Lisbon, a brief look at the main relevant transformations to space and society in recent decades and at the supply and demand sides of the GC market is required. The Portuguese democratic revolution of April 1974 ended the corporative dictatorship that had run the country for almost half a century. Even if that authoritarian regime termed itself the Estado Novo (New State), it was neither modernisation nor values of the Enlightenment that took precedence in its ideological and political programmes or its economic and social policies. The years following regime change were to see a number of deep and rapid social and economic changes (resulting in a general improvement to standards of living) that were to be further amplified and consolidated after 1986 when Portugal became a member of the European Union. Late and unevenly modernised, Portugal has since experienced changes usually identified with postmodern and post-Fordist trends. These help account for the complex current picture of its space and society in particularly in the Lisbon metropolis. In this context, gated communities prove to be paradigmatic.

First, they echo, in differing ways, some of the more recent and important transformations in Portuguese social structure. These include rapid upward social mobility for some sectors of the population, growing poverty, social exclusion (Capucha 1998) and social polarisation (Gaspar, Henriques and Vale 1998); the LMA experiences these to a proportionally greater extent. Throughout the last two decades there has been relative growth in certain segments of the middle and upper middle classes, mainly composed of managers, senior executives, technicians and scientific and intellectual professionals. At the end of the 1980s this corresponded to around 10 per cent of the active population (Peixoto 1999). This reveals the growing importance of academic qualifications to Portuguese social mobility processes (Estanque and Mendes 1997) and certain professions in the tertiary sector as an outcome of the restructuring of the economy and production. It is no coincidence that the main clients of gated housing estates belong to those socio-economic segments, as is indicated by the interviews conducted, observations of characteristics (specifically the portrayed lifestyles and the extolled social images and values) and the prices of GCs (usually much higher than alternatives). The other transformations in social structure should also be taken into account in this context. In fact, phenomena like growing poverty, social exclusion and social polarisation, besides being partially their counterpart, may generate many social tensions, enforce generic feelings of social insecurity and disorganisation, and invite segregation.

Second, GCs are an example of the new *packaged landscapes* (Knox 1992), which include new housing and commercial, work and leisure environments, that during the last decade began visibly to fill the increasingly fragmented body of the Lisbon metropolis. In fact, spatial fragmentation, a trend usually associated with post-Fordist and postmodern transformations (and in particular in the North American metropolis), is also clearly at work in the LMA (Salgueiro 2001), where it encountered a specific terrain. As noted above, due to late modernisation and its consequences, Lisbon, like many other southern European metropolises, including Madrid, displays more 'disorganised' social and spatial patterns than the northern European metropolises. Even if the continental European urbanistic tradition informed its basic modern organisation from the 19th century onwards, the cities of the south demonstrate less well-defined functional and social boundary developments. During the last two decades this inherited pattern has become even more complex as space reflected deep economic and social changes. This was, in turn, magnified by an extremely rapid pace of change, a persistent deficit in public planning and space regulation and the emergence of neo-liberal perspectives fostering market freedom of action.

In terms of the latter, some aspects are worth mentioning when it comes to analysing attitudes held by the Portuguese public authorities toward GCs. During the early 1990s new legal regulations of horizontal ownership and of subdivision and building operations came into force. These regulations included certain clauses that, without ever naming or directly mentioning GCs, clearly focused on serving their institutionalisation and on facilitating, if not encouraging, their development. The municipal authority response to GCs and to their administrative approval has also been generally very favourable due to several factors. GCs, or *condomínios fechados*, are usually envisaged as the merely technical and not as the political. Where they meet all legal and technical requirements they will be approved, as would any other kind of building or subdivision process (no questioning or debate as to its socio-spatial specificity or possible impact takes place). Moreover, municipalities can obtain several practical advantages from approving GCs – the revenue generated by authorising developments, lesser demands on public goods and the prestige that especially the bigger developments are perceived to bestow on municipalities given how they are usually interpreted as 'quality projects'.

Third, *condomínios fechados* developments may be understood as a mirror of the evolution of the Portuguese real-estate sector throughout the last two decades and of its actual structure. To start with, it should be noted that GCs in Portugal were basically driven by the supply side of the respective market. This conclusion is demonstrated both by the interviews conducted and by observing the evolution and dynamics of the LMA real-estate market and its relationship with the processes experienced in introducing and expanding this specific product. Moreover, there is enough evidence to suggest that GCs were introduced in Portugal via Brazil, where they have represented an important reality since the 1980s (Caldeira 1996; Ribeiro 1997). At least 23 of the 97 surveyed GCs in LMA were developed, co-developed or commercialised by firms of Brazilian origin or whose key executives had previously had a professional career there.

Among the agents on the supply side of the GC market we find different profiles. Some of them reflect some of the more important trends in the Portuguese real-estate sector of the last 20 years, such as increasing specialisation, professionalisation (Salgueiro 1994), modernisation, internationalisation and product diversification. However, other developers still retain the traditional business form of the small building firm. This kind of developer is usually associated with very small-scale developments and not usually present within Lisbon's municipal boundaries, where land prices and the nature of the market serve as a deterrent to small-scale developers. Such firms are more at ease in suburban municipalities where they usually face a much more favourable economic and institutional environment.

Finally, LMA GCs reflect the change, (post)modernisation and globalisation of Portuguese cultural patterns, models and values, and of lifestyles, in particular among the middle- and upper middle-class segments of the population. Through an analysis of the symbolic content conveyed in GC planning, architecture, landscaping and promotional texts, certain typical symbolic images were identified (all very similar and quite independent of the development's formal characteristics). Generally these commodified utopias are promoted as 'places' where society, space and time are completely different and apart from (and superior to) the 'normal' or outside world.

Society is usually presented as a world where social relations and social life are warm, authentic and moral, as is idealistically conjured up by the very idea of *community* or *Gemeinschaft* (Tönnies 1979). However, it claims to be simultaneously civilised, respectful of (modern) individual and family privacy and basically rational, neighbourly, convivial and 'clubby' – corresponding to the very logic of *Gesellschaft* (ibid.) or the contractual *association* of mutually interested and socially alike but independent individuals. *Space* is represented as safe, controlled, ordered and beautified, as a *landscape* expurgated of its *dark side* (Barrell 1980) or, as W.J.T. Mitchell (1994: 15) would put it, a *real estate* that became an *ideal estate*. Finally, *time* is a promise of a bright and new future, sometimes with clear references to modernity and newness, perfect for starting over, while also frequently nostalgic of an aristocratic past.

The rise of gated and pseudo-gated communities in Madrid

In Madrid the rise of gated housing estates follows a different logic than the Lisbon case. Explaining such differences involves a short chronology and classification of gated neighbourhoods in Madrid. The development of gated housing estates in Madrid began in the 1950s with isolated settlements on the northern periphery. Fuente del Fresno is probably the first. Originally planned as a weekend and holiday resort some 25 kilometres from the city centre and far from the dusty and loud metropolis, construction began in the latter half of the 1950s. Other private housing estates followed, including for example Ciudalcampo, Ciudad Santo Domingo and Cerro Alarcón (figure 12.4). These communities symbolised a new type of settlement: peripherally located, in many respects independent from the municipalities their terrain belonged to, with an administration of their own and, usually, a transferred responsibility for the formerly public infrastructure such as internal roads, community lighting, water and sewage facilities. This naturally includes gates, walls and a private security system. But unlike US-type gated communities, the common infrastructure in most cases did not pass into the property of the community. Normally, only oral contracts were agreed between the municipalities and the GC administration as to the regulation of responsibility for common infrastructures, and as to the cost of maintenance that has been transferred to the community. In compensation, the neighbourhood gained the tolerance to establish a private security system.

Nowadays, some 35 years on from the establishment of the first large gated housing estates on the periphery of Madrid, this sort of tolerance system as to the gating of what is still officially public property has resulted in many gated communities experiencing complications between public authorities and private community administrations (see below). After Franco's death (1975) and with the transition of Spain from dictatorship to democracy, one of the characteristics of the urban development context has been the weakness of the public authorities. Urban politics and planning have suffered from uncertainty and a lack of self-assertive power caused both by expected changes in legislation and a loss of popular authority. This 'weak state' opened up a vast negotiable terrain between private developers and public representatives. As a consequence, space became much more heterogeneous than was previously the case.[1]

The amplification and the rise of new gated communities was one effect of this transitional phase, perhaps comparable with urban development processes in neo-liberal times. Negotiation and laissez-faire are the keywords characterising the governance system and, consequently, the means of urbanising the periphery from 1975 through to the early 1980s. Many housing estates, mostly

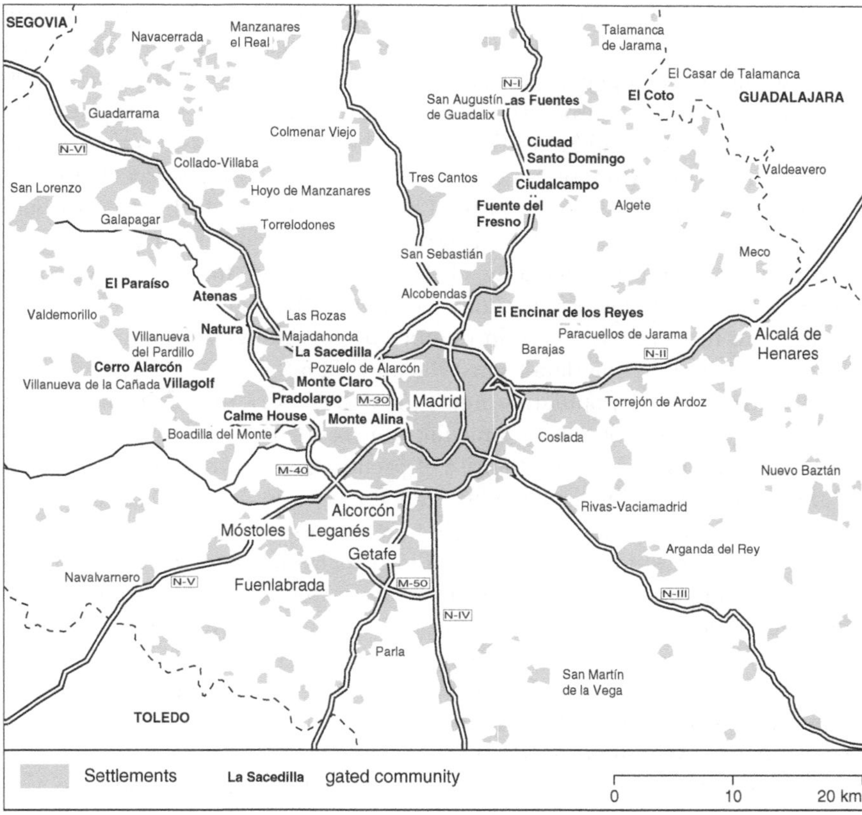

Figure 12.4 Localisation of selected gated and pseudo-gated communities in Greater Madrid. Source: Wehrhahn 2001

non-gated, and predominantly located in the western part of the periphery of Madrid, developed during this period. Gated communities established in those times included, for example, Monte Alina and Monte Claro.

From the mid-1980s onwards, we can identify three different processes:

1 the transformation of single-family estates originally intended as weekend resorts into permanent residence estates;
2 the development of single-family settlements with private security systems of a greater or lesser effectiveness ('pseudo-gated individual housing estates'); and
3 the rise of a rather new form of gated community, the condominium type, usually built as smaller row houses or apartment complexes.

The first process, the transformation of weekend resorts into first-residence estates, occurred especially in the 1980s in conjunction with the economic boom of the latter half of the 1980s. While suburbanisation became a common phenomenon in Spain much later than in western and central European and North American countries, this boom generated a new rise in construction activities not only in general terms on the urban periphery (Estébanez 1990; Wehrhahn 2000), but also within the most attractive gated communities like Ciudalcampo or Ciudad Santo Domingo. The last free plots became occupied during this period. More peripherally located communities, especially those in the north-west and south-west of the Comunidad de Madrid passed through this process only at the end of the 1990s. Examples include some pseudo-gated communities in the municipality of Torrelodones or the gated community Cerro Alarcón in Valdemorillo. Pozo Rivera and Rodríguez Moya (1998: 313) attribute to this process a certain number of (originally) secondary residence estates, using population data from 1986 and 1996.[2]

In Cerro Alarcón population figures rose from 127 to 675 in this period. Most of the new inhabitants became full-time residents.

The second phenomenon, the rise of pseudo-gated communities, occurred in the 1980s and gained more importance a decade later. Suburbanisation, and the new ideal of living in a single-family house with garden, accelerated the urbanisation of peripheral areas. This process has been combined in the case of pseudo-gated communities with the introduction of private security measures. They consist of patrolling security personnel, symbolic gates and/or symbolic entrance controls. Streets and green areas are not private, access is open to everyone, but signs, barriers and security personnel suggest a private atmosphere within the settlement and seek to prohibit non-authorised persons from entrance (Wehrhahn 2001: 108).

Condominium-type estates with 50 to 200 housing units became very common in Madrid in the 1990s, although the first complexes date from a slightly earlier period.[3] Condominiums, defined as smaller housing estates of apartment blocs or row houses, or a combination of both (Glasze 2001, 2003; Wehrhahn 2001, 2003), are well described from very different parts of the world from North America through to developing countries like Brazil, where they fulfil the function of enclaves for the better off. In Madrid, in general, condominiums are supposed to accommodate almost all socio-economic classes, but they are predominantly demanded by members of the middle class who can afford living in a controlled (suburban) environment with some private leisure infrastructure. A special form of this type are small units of row or detached houses with controlled access but with very limited common infrastructures, usually limited to a pool or a playground, and a rather basic internal administration.

Now, in the early years of the 21st century, gated neighbourhoods in Madrid are mostly developed in smaller units, morphologically formed of apartment complexes and rows of houses. A new current phenomenon is the rise of very small, usually luxurious and strictly gated and controlled units within larger settlements. Apartment and individual-house neighbourhoods within the well-known *Urbanización El Encinar de los Reyes* represent an example of this process of separation within an already pseudo-gated settlement.

Explaining Madrid's gated communities

Explanations for the rise of gated neighbourhoods in Madrid will have to take into account the different types of community within their context of emergence and subsequent development. Time and space, societal and economic changes, housing and settlement culture, political and economic institutions and other actors and their interests and strategies have to be incorporated. All these components have to be considered in the light of global and local processes and the categories of demand and supply of new housing forms.

Some factors like the political circumstances, the function of time and general economic development (after Franco) in the process of suburbanisation and the role of public institutions, for example in the tolerance of literally spoken agreements resulting in the illegal closure of settlements, have already been mentioned. Evidently there are strong correlations between phases of economic development and urbanisation (cf. Méndez 1994, 2001) influenced and/or accompanied by political upheavals and changes on the one hand and the rise of gated neighbourhoods on the other. However, gated communities represent only one element of the complex change from industrial to post-industrial (or 'postmodern') cities (Soja 2000), which has at least been proven by the highly fragmented urban periphery of Madrid (Wehrhahn 2000). This metropolitan area can be interpreted as an outstanding European example of the spatial manifestation of economic restructuring, social polarisation and the pluralisation of lifestyles.

The role of politics and planning as well as the attitude of real-estate companies are also important to the supply side of private-neighbourhood analysis, as the example of Lisbon has shown. A detailed discussion of their role in Madrid is provided below. However, the demand side must first be considered as its components are strongly linked to the behaviour of environmental actors like planning institutions and private companies in the real-estate sector.

Differentiating between strictly gated communities and pseudo-gated communities will help us understand why people increasingly move to gated and pseudo-gated neighbourhoods. The effect of the gates may be almost the same, but the reasons for fencing and walling seem to be different. Pseudo-gates are certainly not good as effective entrance controls. This means that a perceived lack of security in the prior location can be excluded as a significant motivation when opting to live behind only symbolic gates. People who are really endangered or who feel at risk due to their profession or lifestyle do not chose pseudo-gated communities unless especially protected within their own property.

Interviews with inhabitants of gated and pseudo-gated communities in Madrid reveal the differences in motivations for choosing a certain type of community (cf. table 12.1 overleaf). The security system is much more important for those living in strictly gated communities. Prestige is always a relevant factor, but slightly more relevant in gated communities of high social status that have effective private security systems at their disposal. But what is interesting

Table 12.1 Motivations for the move to gated and pseudo-gated communities in Madrid.

Motive	*Gated communities*		*Condominiums*		*Pseudo-gated communities*	
	Enquiry A	*Enquiry B*	*Enquiry A*	*Enquiry B*	*Enquiry A*	*Enquiry B*
Security (in general)	++	++	++	++	++	+
Security system	++	++	0	Varying	+	0
Prestige I: image	++	++	++	+	+	0
Prestige II: social level	+	++	+	0	+	n.d.
Community	–	–	–	0	–	–
Leisure infrastructure	0	+	(+)	(+)	0	–
Private governance	n.d.	0	n.d.	–	n.d.	– –

Source: enquiries by Wehrhahn 2000

Notes

- Enquiry A: joint results of a standardised enquiry (open and closed questions)
- Enquiry B: interviews with representatives of gated-neighbourhood administrations, planning institutions and real-estate companies, and with security personnel
- The results of both enquiries have been reassigned in a qualitative procedure to a five-level scale: '++' (very important) to '– –' (very unimportant); the symbols do not represent percentages
- (+) only valid for those communities with leisure infrastructure
- n.d. = no data

is that inhabitants of pseudo-gated communities, with scarce security facilities and rather mediocre green areas, entrance portals and other club goods like sports facilities, do stress the importance of prestige symbols for their motivation to move to this settlement. Elements of gated private neighbourhoods, even if they are neither effective, nor beautiful, nor prestigious, obviously do play an important role in the demand side as they symbolise the new, the different and the distinctive style of housing in postmodern urban society. As a consequence, pseudo-gated communities can be interpreted as proof that the construction of 'new' housing imaginaries has been expanding in Madrid in the last two decades.

The relevance of the variables 'closed community' and 'private governance' reveals a great difference in Spanish gated communities when compared to gated communities in the US as well as in developing countries. The community factor does not play a significant role in any of the considered communities. Community life is usually limited to the necessary meetings and is not classified as an important factor in the decision to live in a specific housing estate. Private governance has never been classified as a relevant factor in decisions as to where to live. It is not the weakness of public government and the feeling that privately managed communities work better that lead to the choice of living in a more or less gated community.

As a consequence, the weak-state hypothesis, mentioned earlier as one possible explanation for the rise of gated communities during the *transición* period following Franco, is only partially valid as an explanation. Inhabitants of all analysed types of gated and pseudo-gated communities in Madrid demonstrated very little interest in administrating the community on their own. Private governance is regarded as merely a necessary evil. The 'weak-state' factor seems to be limited to the aforementioned period of the 1970s and early 1980s. This finding is supported by two processes that are currently ongoing. First, there are some pseudo-gated communities in Madrid which are no longer able to maintain their common infrastructures, especially streets and green areas. Those communities are now forced to negotiate with the public authorities over the return of their 'privatised' infrastructure to the municipality – and, consequently, eliminating barriers and renouncing access controls. A second current process also concerns the tolerance system but from another perspective. In the late 1990s some municipalities, depending on their political position, began to insist on public access to legally public roads, even if there were some existing (oral) contracts from the past as to the private use of such roads or paths. *Cañadas* or cattle paths, for example, always have to be kept open to everybody. Some larger gated communities, such as Ciudad Santo Domingo, are already affected by these trends to open up. Thus, generally speaking, institutions may be changed and this is an example of change running counter to the more common process of an increasing privatisation of public space.

With respect to the aforementioned role of the real-estate sector in Madrid and especially in the periphery of the metropolis, public authorities in many respects can be characterised as relatively weak counterparts to the major development companies and real-estate actors in general. However, the cited case of older and larger gated communities now having to open up their space only shows the potential power of political and planning institutions. Nevertheless, real-estate companies, national and international, currently decide to a large extent the quantity and quality of the urbanisation process in Greater Madrid.

In comparison to Lisbon, such companies rarely advertise the gates and walls. The incidence of terms such as *urbanización cerrada* ('gated settlement'), 'safe', 'security' and related expressions in real estate advertisements is very low. Interviews with representatives of the real-estate sector support this finding. Real-estate marketing strategies refer to leisure infrastructures and other facilities within housing complexes while usually omitting safety attributes such as controlled access and gates. Gates are recognised as prestigious symbols and perhaps as safety features by buyers but it seems to be unnecessary to emphasise such amenities for marketing purposes.

Summarising the role of the real-estate sector, it clearly has a major influence on the periphery of Madrid in general while not determining the proliferation of gated housing estates. As an unproved and weak hypothesis, this may be explained by the large amount of national real-estate companies which may remain attached to a Spanish way and style of settlement and housing culture, at least as regards gating and controlling. A second and more likely hypothesis refers to the Spanish *portero* system (doorman system; Wehrhahn 2003) that may explain the familiarity of Spanish people with control systems within their private housing sphere – unlike central European habits. This presumption may give us some evidence why neither supply- nor demand-side actors emphasise the fortification aspect in the Madrid case study. Such a hypothesis requires verification through further research.

Conclusion

Both the Greater Madrid and Lisbon Metropolitan Area cases at first glance demonstrated various similarities. However, more detailed analysis also revealed significant differences, especially with respect to explanations of the GC phenomenon in the context of societal, institutional and spatial developments in the two metropolitan areas. While gated neighbourhoods both in Portugal and in Spain emerged in the last third of the 20th century, the first settlements in Madrid actually date to the 1950s. In contrast to Lisbon, they were conceived as weekend resorts for the better off and were transformed into permanent resident settlements in the form of large single-family units only over the course of the last two decades. Furthermore, in Madrid it is possible to link different types of gated community development to specific political and economic events or periods.

In Madrid and Lisbon, the observed process of emergence and proliferation of gated neighbourhoods can be connected with a phase of suburbanisation and with the process of postmodernisation in the last decade. In these cities, suburbanisation took root during the late 1960s and the 1970s, and thus rather late when compared to central and northern European countries, due to late industrialisation and modernisation. However, from the second half of the 1980s onwards it gained new impetus and new direction. Huge row-house estates and large apartment complexes for the wider middle class now began to fill those metropolitan areas. In the context of this urban periphery restructuring, and quite unlike the USA in terms of time and quantity, gated and pseudo-gated communities made only a small contribution to suburbanisation in both Spain and Portugal. However, despite this limited quantitative impact, gated neighbourhoods play an important role in terms of the qualitative urban restructuring.

Gated communities significantly increase the fragmentation of urban space and thus testify to a rising socio-economic and socio-cultural heterogenisation and polarisation in space. The latter, separation by gates and walls and by private security employees, as well as by prestigious club goods, must generally be interpreted as a *social distinction* (Bourdieu 1979) manoeuvre. In both Lisbon and Madrid GCs work as a kind of a 'premium' for, and an objective mark of, positive social mobility and thus of social and spatial separation. It does not even particularly matter whether or not they are strictly or pseudo-gated (a case much more frequent in Madrid than in Lisbon) as they can all fulfil, with slight differences, the same general primary objective – symbolic gates becoming a sufficient symbol for individual wishes in a rapidly changing society.

The influence of globalisation and international models of housing seems to be present in both Madrid and Lisbon although our findings are quite heterogeneous. The Lisbon case shows a rather high sensitivity towards more globalised, and specifically North American, ways of living and housing, as was concluded from the analysis carried out of publicity strategies and of built form. In Lisbon real-estate advertising is much more explicit than in Madrid with respect to gates, privacy, security and so on. Moreover, advertisements convey a range of symbolic images presenting gated communities as living paradises. This type of advertising is scarce in Madrid. At the same time, and perhaps suggesting an explanation for cross-metropolitan differences, gated housing estates in Lisbon are frequently marketed by Brazilian-based or Brazilian-related companies, in contrast to Madrid, where national companies are much more important. On the other hand, very often in Madrid the 'Anglo-Saxon' style of row and semi-detached houses, normally non-gated, is obvious and sometimes serves as an explicit marketing component. All in all, our findings for the 'global–local discussion', with respect to their potential influences on gating and fencing, are not exhaustive and consequently are only useful as a starting point for further research.

As for the weak-state hypothesis, we have already referred to certain results. In sharp contrast to some US situations and particularly those of developing countries, Spanish public authority conditions generally do not induce people to close their neighbourhoods for safety reasons or in order to improve the (public) infrastructure. In Lisbon, we find a slightly different situation. Even while this case is also far from being comparable to the US or developing countries, the weak-state hypothesis seems to hold some interest. In fact, the specific balance between private and public agencies in Portugal does seem to favour the development of gated neighbourhoods. Moreover, in the Lisbon case we cannot dismiss the hypothesis that people are also searching for amenities and an improved environment that the public sector cannot fully provide, especially when it comes to amenities that are expensive and relatively uncommon and belong to foreign lifestyle models, such as tennis courts, golf courses and so on.

Moreover, our empirical data clearly prove that private governance is not at all a motivation for people choosing a gated private or privatised community for their place of residence. In both the Lisbon and Madrid cases, people seem just to accept it as a 'necessary evil' and not as an option. Although in some individual cases in the Madrid region some interview partners declared that a separation of their community from the municipality might be an option, this cannot be taken as a general rule and has always been connected with a current dispute with municipal authorities and/or politicians. And, in contrast with some US or Brazilian cases, in Spain and Portugal the secession of a part of the municipality would indeed prove to be very difficult to underpin legally.

Another interesting aspect discussed in the literature about private neighbourhoods is the importance of the community feeling or the 'search for community'. Again in contrast to the North American debate, neither private administration, in the sense of self-determination, nor the desire to build a community, in the sense of *Gemeinschaft* can truly be found in our case studies. The members of gated and even more of pseudo-gated communities in Madrid live together in neighbourhoods largely lacking in effective or strategic objectives and intentions. In Lisbon this finding can be widely corroborated. Demand for this type of settlement is principally generated by status and prestige and to a lesser extent the wish to exclude non-members. However, the pursuit of a relatively homogeneous neighbourhood – which is generally a common pursuit – does not lead to rules for the community that exclude certain population groups by definition. Regulations of gated communities in Madrid and Lisbon are usually limited to technical aspects. The choice of new members, e.g. by age, is not possible. It is not known if there have been any attempts to regulate this kind of membership selection but European states' laws are very sensitive to discrimination, something that Davis (1990) deplores about the US where legislation is quite different. This is a matter that deserves further research as it does seem to correspond to one of the more important differences between US gated communities and those that have emerged in Europe.

Notes

1 Urban planning has not been as regulated as one might expect by a state with the authoritarian traditions of Franco's times (cf. Estébanez 1990; Canosa 1995).
2 But, like other Spanish authors, Pozo Rivera and Rodríguez Moya (1998) do not mention the gated aspect when analysing population processes in the urban periphery.
3 For instance, the gated apartment complex La Sacedilla in the 1970s.

Bibliography

Barrell, J. (1980) *The Dark Side of the Landscape: The Rural Poor in English Paintings, 1730–1840*, Cambridge: Cambridge University Press.

Blakely, E.J. and Snyder, M.G. (1997) *Fortress America: Gated Communities in the United States*, Washington, DC and Cambridge, MA: The Brookings Institution.

—— (1998) 'Forting up: Gated communities in the United States', *Journal of Architectural Planning Research*, 15: 61–72.

Bourdieu, P. (1979) *La Distinction. Critique sociale du jugement*, Paris: Éditions de minuit.

Caldeira, T. (1996) 'Building up walls: The new pattern of spatial segregation in São Paulo', *International Social Science Journal*, 147: 55–66.

Canosa, E. (1995) *La promoción inmobiliaria en la periferia noreste de Madrid*, Madrid: UAM.

Capucha, L. (1998) 'Pobreza, Exclusão Social e Marginalidades', in J.M. Viegas and A.F Costa (eds) *Portugal, que Modernidade?*, Oeiras: Celta: 209–42.

Davis, M. (1990) *City of Quartz: Excavating the Future in Los Angeles*, London: Verso.

Estanque, E. and Mendes, J.M. (1997) *Estrutura de Classes e Trajectórias de Classe em Portugal*, Coimbra: CES/FEUC.

Estébanez, J. (ed.) (1990) *Madrid, presente y futuro*, Torrejón de Ardoz: Akal.

Gaspar, J., Henriques, E.B. and Vale, M. (1998) 'Economic restructuring, social recomposition and recent urban changes in Portugal', *GeoJournal*, 46: 63–76.

Glasze, G. (2001) 'Geschlossene Wohnkomplexe (gated communities): "Enklaven des Wohlbefindens" in der wirtschaftsliberalen Stadt', in H. Roggenthin (ed.) *Stadt – der Lebensraum der Zukunft?*, Mainz: Institute of Geography, Mainz University: 39–55 (Mainzer Kontaktstudium Geographie 7).

—— (2003) *Die fragmentierte Stadt. Ursachen und Folgen bewachter Wohnkomplexe im Libanon*, Opladen: Leske and Budrich (Stadtforschung aktuell 89).

INE (Instituto Nacional de Estatística) (1999) *Tipologia Socio-Económica da Área Metropolitana de Lisboa*, Lisboa: INE.

Knox, P. (1992) 'The packaged landscapes of post-suburban America', in J.W.R. Whitehand and P.J. Larkham (eds) *Urban Landscapes: International Perspectives*, London and New York: Routledge: 207–26.

Lash, S. and Urry, J. (1994) *Economies of Signs and Space*, London: Sage.

Luymes, D. (1997) 'The fortification of suburbia: Investigating the rise of enclave communities', *Landscape and Urban Planning*, 39: 187–203.

McKenzie, E. (1994) *Privatopia: Homeowner Associations and the Rise of Residential Private Government*, New Haven, CT: Yale University Press.

Malheiros, J. (2002) 'Ethni-cities: Residential patterns in the northern European and Mediterranean metropolises – implications for policy design', *International Journal of Population Geography*, 8: 107–34.

Méndez, R. (1994) 'Crecimiento periférico y reorganización del modelo metropolitano en la Comunidad de Madrid', *Economía y Sociedad*, 10: 149–73.

—— (2001) 'Transformaciones económicas y reorganización territorial en la región metropolitana de Madrid', *Revista eure*, 27 (80): 141–61.

Mitchell, W.J.T. (ed.) (1994) *Landscape and Power*, Chicago, IL, and London: The University of Chicago Press.

Peixoto, J. (1999) *A Mobilidade Internacional de Quadros*, Oeiras: Celta.

Pozo Rivera, E. and Rodríguez Moya, J.M. (1998) 'La evolución de la población en la Comunidad de Madrid (1991–1996)', *Anales de Geografía de la Universidad Complutense*, 18: 299–316.

Raposo, R. (2002) 'Novas Paisagens: A Produção Social de Condomínios Fechados na Área Metropolitana de Lisboa', unpublished Ph.D. dissertation, Lisbon: Instituto Superior de Economia e Gestão, UTL.

—— (2003) 'New landscapes: Gated housing estates in the Lisbon Metropolitan Area', *Geographica Helvetica*, 58 (4): 293–301.

Ribeiro, L.C.Q. (1997) *Dos Cortiços aos Condomínios Fechados*, Rio de Janeiro: Civilização Brasileira.

Salgueiro, T.B. (1994) 'Novos Produtos Imobiliários e Reestruturação Urbana', *Finisterra*, 29 (57): 79–101.

—— (ed.) (1997) *Internacionalização, Reestruturação Económica e Produção de Novas Marginalidades na Região de Lisboa*, Lisboa: JNICT/DGOTDU.

—— (2001) *Lisboa, Periferia e Centralidades*, Oeiras: Celta.

Soja, E.W. (2000) *Postmetropolis: Critical Studies of Cities and Regions*, Oxford: Blackwell.

Tönnies, F. (1979) *Gemeinschaft und Gesellschaft: Grundbegriffe der reinen Soziologie*, Darmstadt: Wissenschaftliche Buchgesellschaft.

Wehrhahn, R. (2000) 'Zur Peripherie postmoderner Metropolen: Periurbanisierung, Fragmentierung und Polarisierung, dargestellt am Beispiel Madrid', *Erdkunde*, 54 (3): 221–37.

—— (2001) 'Gated communities in der postmodernen Stadt, untersucht am Beispiel Madrid', unpublished postdoctoral thesis, Kiel: CAU.

—— (2003) 'Gated communities in Madrid: Zur Funktion von Mauern im europäischen Kontext', *Geographica Helvetica*, 58 (4): 302–13.

13

THE RISE OF PRIVATE RESIDENTIAL NEIGHBOURHOODS IN ENGLAND AND NEW ZEALAND

Sarah Blandy, Jennifer Dixon, Ann Dupuis and David Parsons

Abstract

The different external policy and regulatory frameworks in New Zealand and England might be expected to impact differently on the growth of private residential neighbourhoods. This chapter explores the legal and policy background in each country, as well as the evolution of the internal governance techniques, for this type of development. Despite the disparity in contexts, empirical studies in New Zealand and in England indicate that residents in private neighbourhoods express similar concerns about their management arrangements. The chapter concludes that the planning systems in both countries need attention, as do issues of internal governance, and that the roles of developers and professional management agents should be scrutinised, thus allowing greater power in decision-making by residents.

Introduction

The legal and regulatory frameworks in place at the beginning of the 21st century, in both England and New Zealand, seem to be facilitating the growth of private residential neighbourhoods. The policy objectives of encouraging high-density, inner-city housing developments are common to both countries, although deriving from very different cultural backgrounds and policy concerns. Yet in neither country is the growth of private residential neighbourhoods a policy objective, and in many ways such developments would seem inimical to stated government aims.

This chapter draws on empirical research undertaken in Sheffield, England and Auckland, New Zealand. The Sheffield research follows the development of Nether Edge, a local gated community. The Auckland research reports on a study of Ambrico Place, a medium-density housing development, and subsequent research into bodies corporate, the legal entities that govern these developments (Dixon *et al.* 2001; Dupuis *et al.* 2002). The chapter explores the idea that similar issues concerning private residential neighbourhoods can arise in contrasting circumstances. These issues can be analysed at two levels. The external regulatory environment, including planning law, government policy and property law, provides the broad framework for the emergence of gated and other common-interest developments (CIDs). The second, more micro level, concerns the mechanisms for the internal governance of private residential neighbourhoods.

A comparison of the legal structures in each country which support the formation of privatised local governance, in both exclusive and more mixed neighbourhoods, highlights the inherent problems of these structures. Interview data with relevant stakeholders confirms these problems. Residents' perceptions of their role in relation to governance of their community are also examined. Their sense of powerlessness contrasts strongly with the intended benefits of participation in resident-run developments. The similarities and differences between English freeholder management companies, or the new commonhold associations, and bodies corporate, their New Zealand counterparts, provide valuable points of comparison.

Planning and policy frameworks

England has a highly regulated system of land-use control which has evolved from the Town and Country Planning Act 1947 into today's system through successive legislative reforms. The planning system is administered at a national level by the Office of the Deputy Prime Minister (ODPM). The duty of the Secretary of State is to secure consistency and continuity in the framing of a national policy regarding the use and development of land.

However, there are some contradictions and tensions in central government guidance on housing. On the one hand the government appears to encourage inclusive communities of mixed social types and housing tenures. The website of the ODPM displays the phrase 'Thriving, inclusive and sustainable communities in all regions'. Planning Policy Guidance 3 (PPG3) encourages high-density developments on brownfield sites, and states, 'Local planning authorities should encourage the development of mixed and balanced communities' (Department for the Environment, Transport and the Regions 2000). On the other hand it refers readers to the Home Office-approved guidance in Secured by Design (ACPO CPI Ltd 2003), which recommends, 'creating real or symbolic barriers for an estate; . . . restricting access for the public through the estate to as few routes as possible'.

These national intentions are translated into policy at the regional level by government offices for the regions (GORs), administrative outposts of four central government departments under the control of a senior regional director. Each GOR draws up regional spatial strategy (RSS) 'to provide a broad, long term strategy for the economy, housing, transport and the environment of the region' (Government Office for Yorkshire and the Humber 2001). Sheffield is within the Yorkshire and Humber Region. The housing policies for this region suggest that 770 houses should be built in Sheffield for each year of the plan period between 1998 and 2016, of which 68 per cent should be on brownfield sites. The plan points out that a significant proportion of projected growth is likely to be among households that will not be able to afford housing in the open market and estimates that affordable dwellings should provide 27 per cent of the total requirement. It recognises that this will differ according to market conditions in different areas and suggests that 'targets for the provision of affordable housing in each district . . . should be developed through housing needs assessments'.

Within this national and regional framework, planning policies are drawn up and set out in local development frameworks by local planning authorities such as Sheffield City Council. Decisions on proposals for new development are taken through the development control process, whereby individual planning proposals are submitted to council officers and decided upon by locally elected members. Local authorities have a large degree of discretion in dealing with applications, which must be decided upon in accordance with their plan unless 'material considerations' indicate otherwise. If the plan does not contain a relevant policy, 'material considerations' will be paramount. These include general principles and policy guidance. Sheffield City Council had no policy on gated communities in their plan (few authorities currently do; see Atkinson *et al.* 2004). The Nether Edge site was developed as an exclusive, single-class, single-tenure gated community, despite this being apparently contrary to national and regional guidance. The process of obtaining planning permission for Nether Edge was investigated by interviewing two officers in the local planning department of Sheffield City Council and two officers in the highways department.

While it would seem that the national guidance on inclusive mixed communities could have been taken into account as a material consideration, in this case it was not, and although PPG 3 is explicit about encouraging high-density schemes, it contains no tools to consider social exclusion. The planning brief prepared by Sheffield City Council for the old hospital site was a material consideration. This brief suggested that 20 per cent of the housing should be affordable for people on low incomes, preferably in the form of socially rented housing rather than low-cost properties for sale (Sheffield City Council 1997, para. 3.6.4). However, in the event the council accepted a commuted sum from the developer and allowed the whole development to be built for private sale rather than insisting on a degree of mixed tenure. In order to develop and market a gated community, the developer wanted the internal roadways to remain private, and found a rather surprising ally in the highways department. The recommendation in Design Bulletin 32, issued by the government under the Highways Act 1980 (Department of the Environment and Department of Transport 1992), is that highways serving more than five houses should be adopted by the local authority (i.e. should be publicly accessible, built to adequate standards and maintained by the local authority). Although this guidance was reflected in Sheffield's own development plan, council highway engineers felt that they had insufficient power to impose a roads adoption requirement for the Nether Edge development. Once this had been conceded, the gates which exclude non-residents were welcomed by the highway department as a sign that this was a private development with responsibility for maintaining its own roads.

Compared with England, New Zealand has a less prescriptive and more devolved planning system with little direct guidance from central government. At the national level, the Resource Management Act (RMA) 1991 is the main statute under which planning policies are formulated by regional and local government. The assessment of environmental effects forms the cornerstone of the act, the purpose of which is 'the promotion of sustainable management of physical and natural resources' (section 5 (1)). The RMA establishes some broad principles within which councils are free to develop their own approach to sustainable management on the basis of avoiding or mitigating adverse environmental effects. This has led to the development of an 'effects-based' planning system, with an emphasis on the management of environmental effects rather than the prescription of activities as occurred under the previous town and country planning regime.

Implementation of the act rests largely in the hands of local (73 district and city councils) and regional authorities (12 regional councils and four unitary authorities). Regional councils are responsible for activities such as managing water, soil, air and coasts; mitigating natural hazards and hazardous wastes; managing discharge of contaminants; and land transport planning. District and city councils are responsible for district planning, noise control, hazard mitigation and land subdivision, along with a wide range of other regulatory and discretionary functions.

The RMA provides a hierarchy of plans where lower-level plans are not to be inconsistent with higher-level plans, hence an integrated system of policy statements and plans govern resource development. These include provision for national and regional policy statements, which set a statutory framework within which regional and district plans are prepared. While the act requires the preparation of a coastal policy statement, other national policy statements can be developed at the discretion of central government. To date, only the coastal policy statement has been prepared. Regional councils are required to prepare regional policy statements which provide an overview of resource management issues in their region. These, in turn, govern the administration of resource consents covering land, water, air and coastal management. Regional plans, with the exception of regional coastal plans, are not mandatory but can be prepared on topics considered appropriate by regional councils. District plans, a mandatory requirement for district and city councils, deal with land-use planning, noise and subdivision controls. In addition, an increasing amount of non-statutory planning activity is now occurring at local and regional levels outside the context of the RMA. For example, within local government, there has been a proliferation of instruments such as growth strategies, strategic plans, structure plans and design guidelines, all of which have policy implications for planning.

Some of these are prepared under the Local Government Act 2002 and some initially outside any existing legislation.

The Auckland region, which has over 30 per cent of New Zealand's population, comprises four cities (Auckland, North Shore, Waitakere and Manukau), and three districts (Rodney, Papakura and Franklin). A particular set of circumstances in the Auckland region led to the development of a coordinated policy and planning framework which has influenced the development of Ambrico Place, located in Waitakere City. Faced with a rapidly growing population and pressing environmental problems, Auckland politicians established the Regional Growth Forum as a voluntary initiative and produced the Auckland Regional Growth Strategy, setting policies in place for the next 50 years (Auckland Regional Growth Forum 1999). A critical element of the strategy is residential intensification. It is envisaged that by 2050 between 25 and 30 per cent of people will live in multi-unit housing in the region, much of which will be constructed in the existing metropolitan area. The strategy has guided the development of the regional policy statement and the seven district and city plans in respect of managing urban growth.

At the local level, the development of the Ambrico Place site was influenced by the Waitakere City Plan (1995), which in fact preceded the regional growth strategy. In particular, the plan provided greater opportunities for intensive housing, taking on board new urbanist principles of more compact residential arrangements and use of public transport. The development of eight separate housing projects in Ambrico Place occurred relatively quickly on the site, following notification of the plan. Council staff also drew on non-mandatory design guidelines to guide decision-making in relation to each of the housing projects. While the Waitakere Plan has been widely regarded as a good example of an effects-based plan, the development of the Ambrico Place site revealed some gaps in this approach to policy-making. In particular, it highlighted the need for more targeted intervention by the council in respect of the development of major sites for urban housing (Dixon and Dupuis 2003). In this instance, there was no requirement for an overall site development plan. Hence there was significant variation with respect to housing quality, layout and design of the eight discrete housing projects.

Internal governance

In America there are three legal frameworks for gated communities, which are often used as models against which those developed in other countries can be compared. In condominiums, each homeowner has freehold title to their own residential unit and in addition an indivisible title, in common with all the other owners, to the common areas (roads, leisure facilities, green spaces). Housing cooperatives take the form of a legal entity which owns the whole property; individual residents are members of the cooperative and as such are granted the right of use of a residential unit and all the common areas and facilities. Common-interest developments (CIDs) are the most common type in

America, where each homeowner has freehold title to their own residential unit, and automatic membership of the homeowners' association which owns the common areas.

Both Ambrico Place and Nether Edge conform most closely to the third type outlined. Under English property law positive covenants cannot be enforced against freehold owners (except against the first purchaser). The residents of Nether Edge own their flats or houses on leasehold tenure. There is no legislative framework for this type of legal arrangement, and no standard clauses which must be adopted. On purchase, each resident automatically becomes a shareholder member of the management company, which is governed by the Companies Acts 1985 and 1989, and which must be registered with the Registrar of Companies. However, apart from conforming to very general regulations concerning filing accounts and so on, the company and others like it can adopt any internal rules they wish.

The developer of Nether Edge (a national company based in Sheffield) invested £30 million in the 4.22-hectare (10.43-acre) site, for which planning permission was granted by Sheffield City Council at the beginning of 2000. Work started immediately on the first phase of construction and the first of 180 dwellings was sold in spring 2001. The development is still under construction, although more than half the dwellings are now occupied. It provides a fascinating study of the stage in a residential development when several interests exist simultaneously: those of the developer, of the residents (individually and collectively), and of a professional property managing agent. The documents which make up the legal framework for the Nether Edge gated community reveal the current balance of power between these different interests at this point, as well as indications of how this may change when the development is completed.

Purchasers are asked to sign lengthy and complex legal documents. There are three parties to the seven-schedule, 24-page lease: the freeholder (developer), the management company and the leaseholder. Each party covenants with the other(s) in a web of mutually enforceable obligations. For example, the leaseholder covenants to pay the estate charge and building services charge, and to comply with 23 restrictive covenants. These include painting the premises internally at least every seven years, cleaning the windows every four weeks, not erecting any structures without prior consent, not using the premises other than for residential purposes, not carrying out car maintenance on the estate, not parking a caravan on the estate and not hanging laundry outside the premises. New purchasers also sign up to membership of the Nether Edge Management Company Limited, a private company formed in May 2001.

Standard clauses in the memorandum and articles give this company wide powers to 'own, manage and administer the freehold property together with the buildings to be erected thereon (the "Common Areas")'. At this stage the management company does not own any freehold property. The shares in the company comprise one 'A' share for the developer and one 'B' share for each dwelling, not to be disposed of during ownership of the unit. Each company member has one vote, except that the 'A' share carries one more vote than is necessary to carry or defeat any resolution at any meeting of the company. The 'A' shareholder constitutes a quorum; but apart from this the company can decide its own constitution in terms of number of directors, and can make its own by-laws. As well as purchasing one 'B' share, each purchaser signs a consent form to act as director or secretary. The first company secretary of the Nether Edge company was another private company, associated with the solicitors who drew up the legal documents, and the first director was an employee of the developer.

Many English property developers use the legal services offered by professional property management companies for setting up the legal framework for gated communities. These companies offer services to developers such as lease creation, formation of residents' management companies, company secretarial services, comprehensive legal documentation and standard model management agreements. Often they are also in a position to control the way in which the management of a development is set up, and to ensure that their role is preserved in the future running of the estate (see Atkinson *et al.* 2004). While the Nether Edge developer employed a firm of solicitors to draft the legal documents, they chose a managing agent with which they have had a long association, although its main offices are in Worcester, which is over a hundred miles from Sheffield.

The managing agent started work at Nether Edge in spring 2002, carrying out the obligations to maintain the estate and its common areas on behalf of the management company, in which currently the developer and all leaseholders are shareholders. Once all the units on the development have been sold, the usual procedure (expected to be followed at Nether Edge in 2005) is that the developer will offer the freehold of the whole site to the management company for a nominal sum. At the same time, the managing agent will write to all the leaseholders inviting them to stand as directors or secretaries of the Nether Edge Management Company. If insufficient numbers are forthcoming, then the managing agent has the right to appoint directors as it sees fit, usually its own employees. In due course the developer will give up the 'A' share in the company. It will then be open to the residents' managing company to vote to employ a different managing agent, although findings from the national study indicate that this is a difficult process (Atkinson *et al.* 2004).

The developer of Nether Edge explained in interview that

> we employ a company who are the managing agents, but they work for the management company. It just so happens that until the last property is sold, plus six months, we control the management company and so we elect to use this particular managing agent. We have, as most owners of developments have, what we call a 'golden share' which gives us command and control until about six months after [the transfer of the freehold to the residents' management company].

There was a sharp contrast between the understanding expressed by the developer of the legal and management framework for Nether Edge, and the lack of knowledge revealed in interviews with the residents. One resident explained that their solicitor 'did tell us that the management committee would be mainly elected from the community and that maybe a company would take control over it and run it on a day-to-day basis'. Another resident recalled that 'we had to sign to agree to be secretary or director or something like that'. Others were even less certain about their control of the development: 'We're not sure about, any of us, how that will materialise. We're assuming that it will be facilitated somehow.' The interviewer explained to each interviewee that the residents' management company would take over the freehold, eliciting responses such as 'Oh, so that is how it will get taken on board', and 'Right, I didn't know that, actually'. One interviewee asked, 'How would the residents' committee be formed?' The developer's honest explanation was: 'You tend to watch how it's beginning to form, and then we cultivate the right people to lead it, which is in our best interests, but also hopefully in the residents' best interests'.

The residents' expectations of the managing agent were optimistic: 'All of the front gardens and things will be maintained; we have full rights on the back. Things like window cleaning are thrown in.' In fact the lease requires the residents to clean their own windows. Early on, the managing agents took on a dispute-resolution role, rather than the residents sorting problems out between themselves. One resident gave this example:

> The person at the front is having a little bit of a dispute with the person at the back about cats, and she has not directed her complaint at all at the person who actually owns the cat, she's directed it to the management company . . . I mean, they are in Worcester which is a bit of a problem.

Residents expressed some uncertainty about how, or whether, covenants would be enforced. One suggested that 'if there were certain gardens that became very untidy, then I think again that would be for the management committee'. Another felt, 'a lot of the things that they've actually put in there [i.e. in the lease], people are already going against and nobody's saying or doing anything'.

The issues of residents' lack of knowledge about the legal structure of their development, and the potential for problems with the managing agent, are both borne out in other research findings. A larger-scale national survey of English gated communities has shown that while a variety of different legal forms are adopted by developers' lawyers, confusion occurs whichever model is chosen, over who really manages the gated community: the developer, the professional managing agent, the residents or an uneasy mixture of some or all of these (see Atkinson *et al.* 2004). None of the models is easy for a layperson to understand. The national study found that sales agents were not able to provide detailed information about how properties in gated communities are owned and managed, so it is up to the individual purchaser's solicitor to explain the structure and the rights and obligations which accompany it. Judging by the lack of knowledge among Nether Edge residents, this rarely happens effectively.

Each legal model provides for a two-stage legal framework; the change coming at the point when, or soon after, the last unit is sold. Initially, full control is in the hands of the developer, as freehold owner of the whole site. Whether, and how, that control is handed over is very much up to the developer. The all-important issue of how the development is managed is also up to the developer, and in most models a professional managing agent is given a degree of control by the developer while the site is being completed. On completion, in some legal models the professional managing agent acquires the freehold; in others their employees are already running the residents' management company when it acquires the freehold.

Concern over the performance of, and control over, managing agents, has a history in England which is closely connected to the long struggle for leasehold reform (Blandy and Robinson 2001). It should be noted here that most leaseholders have no control over their freeholder's choice of managing agent, whereas the Nether Edge residents theoretically will do so, as the developer's freehold interest and 'A' share will both come to an end. Following the 1998 government consultation on proposals for leasehold reform, 96 per cent of responses favoured additional controls over managing agents (Department for the Environment, Transport and the Regions 1999). Yet the only recent move towards regulating managing agents has been the approval by the Secretary of State of the Royal Institution of Chartered Surveyors' *Code of Practice* (1997). In a recent consultation paper on residential leasehold management, the government acknowledged that 'under existing law, anyone can manage leasehold property. There is no legal requirement for managers to demonstrate that they possess certain knowledge or skills. Nor does it matter whether they have a poor track record' (Department for Transport, Local Government and the Regions 2002: Introduction, para. 1). No analysis of the responses to this consultation has yet been reported.

The Commonhold and Leasehold Reform Act 2002 offered an opportunity to legislate for regulation or licensing of managing agents, but does not address these issues. The act does introduce commonhold, potentially an alternative legal framework for gated developments, although it should be noted that developers will still have the choice between using commonhold or the existing legal frameworks. This new form of land-ownership brings English law into line with other countries, for example the New Zealand Unit Titles Act, the Australian strata titles, and the American CIDs. In England, a commonhold association must take the form of a company governed by the Companies Acts. Individual unit-holders will own their own unit as freeholders, and collectively own the common parts of the land and property on freehold, as members of the commonhold association. The legislation provides that statutorily defined rights and obligations, and communal management, will attach to each unit. The issue of enforcement of positive covenants is dealt with by making all unit-holders subject to the same rights and obligations, and providing that these will apply to all future unit-holders, although the obligations cannot be enforced by foreclosure. It will still be open to developers to appoint employees of professional management companies as directors of commonhold associations, who might well become influential because of their expertise and early appointment. The commonhold rules allow a developer to appoint, and remove, up to one-quarter of the total number of directors while the developer retains ownership of one-quarter of the units. Directors need not be members of the commonhold association, but may be employed by it. It cannot therefore be assumed that, even as freeholders from the outset, the residents of a commonhold development would be in a better position to exert control than when leaseholders in resident management companies.

The standardised commonhold model documents, in contrast with individually drafted leasehold schemes, are intended to make the legal framework simpler, and cheaper to implement. However, lawyers for developers and property management companies have not been positive about using the commonhold structure (Atkinson *et al.* 2004). This was a typical response, from a lawyer for a developer: 'I accept that rent charges and the positive covenant system are archaic, but at present they seem the best option'. They are the best option because they allow the developer to retain control until at least the end of the development stage, in contrast to the commonhold framework where the association takes over as soon as the developer sells the first unit and the developer only has control of up to one-quarter of the directorate. Commonhold also poses problems for larger developments as it does not easily allow for phased developments, or for changes to be made in the number of units, once commonhold ownership has come into effect. It is likely therefore that the initial take-up of commonhold will be low, and probably limited to smaller developments. The consequences for management by residents will only emerge over time.

Unlike the Nether Edge development, which is gated and has been developed as a coherent entity, Ambrico Place in New Zealand comprises eight separate terraced housing developments of varying quality, that were built at different times by different developers. With approximately 300 units, by Auckland standards Ambrico Place is large. Early on it was evident that there were major issues surrounding the planning processes for Ambrico Place. The non-notification of applications for resource consents (that is, applications for development proposals lodged with councils) and the speed at which the eight Ambrico Place developments were constructed, meant that the first groups of incoming residents were unaware of other pending developments on the site. The largest and highest-quality development, Tuscany Towers, was first on site, followed subsequently by developments of lesser quality. This led to criticism of the way in which the council had handled development of the overall site and concerns by the Tuscany Towers developer and residents regarding the possible impact on property prices (Dixon *et al.* 2001).

Like most Auckland intensive housing developments, Ambrico Place comes under the body corporate legislation within the Unit Titles Act, 1972 (UTA). Under the UTA land and buildings are subdivided into units and common property by means of the deposit of a unit plan under the Land Transfer Act. The body corporate is established at the same time as the unit plan is legally deposited. It is responsible for the maintenance of common property, for setting levies on owners for their financial contribution, for the arrangement of insurance and valuation, for preparation of annual accounts and for conduct of annual general meetings. The rules of a body corporate are set out in the second and third schedules of the UTA, but can be amended at the time the body corporate is established. Usually a body corporate is managed by a specialised company. The second schedule of the act makes provision for the body corporate to appoint a secretary, although it is also possible for an owners' committee to take on this role (Dupuis *et al.* 2002).

The Ambrico Place research identified several issues concerning the operation of bodies corporate (Dixon *et al.* 2001). Seventy-two per cent of respondents interviewed reported dissatisfaction with their body corporate, frequently citing poor communication between body corporate managers and owners. Residents' suggestions for the improved operation of bodies corporate included replacement of the body corporate agent, the need for a stronger presence from the body corporate, greater owner autonomy and the provision of more information and greater transparency in management. In the case of tenanted property, owners are expected to inform their tenants of body corporate rules. Forty-five per cent of the units in the Ambrico Place Research Project were rented. Fewer than one-third of tenants knew about the existence of the body corporate that managed their unit. As a consequence, they had no knowledge of body corporate rules and so, unknowingly, breached rules relating to issues such as parking and rubbish disposal. Body corporate managers who were interviewed identified the need for active owners' committees, ways of addressing language barriers, the employment of building managers, closer attention to design details and the importance of the development of long-term maintenance plans.

Bodies corporate in New Zealand are strongly shaped by the powers of the developer and by body corporate management companies. During the planning phase developers may have contact with one or more body corporate management company for the purpose of contracting a company to oversee the body corporate once it has been established. At this time, the body corporate management company may advise the developer on what second- and third-schedule rules to include. It may also (and without charge) obtain estimates of valuation and insurance costs; prepare a budget for consideration, including an estimated share per unit of body corporate costs; and arrange copies of the budget for marketing purposes.

In effect, and in law, owners are the body corporate. Initially, this usually means that the developer (or development company), as the first registered proprietor of all units, is the body corporate. As the developer sells the units the new proprietors also become part of the body corporate. Once all the units are sold the developer is no longer part of the body corporate. What this means, however, is that subsequent unit owners do not set the rules or appoint the body corporate management company. Indeed, it may be some time before an annual general meeting is held, where owners can voice their concerns. In this intervening period, developers can take advantage of the lack of a formal forum to consolidate contractual arrangements and other measures advantageous to themselves. For example, the developer may still own a number of units and could block any moves by other owners to amend rules.

Numerous issues were raised by research participants over body corporate rules. Interview data highlighted the considerable variation in both attitudes and practices of developers with regard to the setting and content of body corporate rules. While some developers established broad rules which future owners could amend as necessary, others sought to entrench particular rules so that they became impossible to change, an action that, if tested in court, could be deemed *ultra vires*. One example is where a rule is written to appoint a particular company to act as the body corporate secretary.

Problems also arose in instances where body corporate rules were written in a way that made it almost impossible for the body corporate to change the management company, initially engaged by the developer, or advantaged management companies over residents and owners. A spokesperson from a body corporate management company said,

> they'll put things in there like 'the Body Corporate Secretary will be Joe Bloggs Ltd for the next ten years'. Well that's taking their [owners'] right of choice away for the next ten years. There are some very unhappy people out there that are locked into a body corporate secretary that they just can't seem to get away from.

In a similar vein an owners' committee spokesperson complained about the lack of power of his body corporate and the rules that made the secretary's dismissal almost impossible. Although the Unit Titles Act was introduced more than 30 years ago, it has remained substantially unchanged since then. The recent explosion of intensive housing has meant that practices, such as those described above, are relatively new and untested.

Second-schedule rules can be changed in two ways: a unanimous vote of owners, or support by at least 80 per cent of owners before the matter can be taken to the High Court. Taking this action is costly, difficult and time-consuming. It is clear, therefore, that decisions made by developers when the body corporate is established are pivotal for the long-term operation of the body corporate. Yet the developer has very little, if any, accountability once the units are all sold. On occasion, the developer has no liability as the development company has been wound up and thus the developer has moved on by the time body corporate problems arise. This suggests that at the earliest opportunity the first group of new owners should have the chance to review the rules and contractual arrangements initially set in place. In situations of conflict, the difficulty with changing second-schedule rules can mean that owners feel disempowered and frustrated.

The key groups with various responsibilities under the UTA are developers, body corporate secretaries or management companies, owners' committees, property management companies and on- and off-site managers. The extent to which these sets of people have relationships and the nature of those relationships vary over time, according to whether the development is in its planning, construction or occupation phase. Although the developer is usually only involved until the units are sold, there is regular contact between most of the other groups, once the housing is occupied.

The body corporate research identified significant problems in the day-to-day relationships between stakeholder groups. Many of these problems could be attributed to outdated and inadequate legislation. However, there is no doubt that some parties had taken advantage of legislative loopholes, to the extent that some research participants commented on the unethical practices they were aware of. For example, one developer reported, 'The reason why some developers have had, and some still do the ten-year thing is because some body corporate managers are willing to pay for it. They're willing to give the developer some money for a ten-year contract'.

Numerous conflict-of-interest situations were also reported. These occurred in the relationships between property management companies and developers and owners in the following areas. First, property management companies may receive anywhere between a one- and a ten-year contract from a developer to manage a housing development. Once contracted, however, it is the body corporate which pays the property management company from body corporate levies, despite having had no choice with respect to any of the decision-making regarding the appointment of this company. Second, the property management company and the development company may be financially linked (e.g. have the same director). This sets up an immediate conflict of interest in that the ability for both parties to act impartially may be undermined, thus operating against the interests of the owners. Third, a property management company sometimes buys a unit in a development to house an on-site manager. The body corporate pays rent to the property management company for this unit, as well as the wages of the on-site manager. A conflict of interest may then arise as the property management company is, in turn, a member of the body corporate with full voting rights.

It was also evident that there was considerable unevenness in the quality of service offered by body corporate management companies. While some body corporate management companies had excellent reputations, several owners interviewed cited cases where relationships had broken down with their body corporate management company. Developers and lawyers were also concerned about the officious behaviour of some body corporate management companies and their appointed secretaries. There was general agreement on the need for greater regulation of body corporate management companies. For example, one lawyer said:

> A fair amount of my work involves body corporate secretaries who are doing less than a good job . . . and they can get away with it . . . Some of the things that they do are quite shocking – failing to hold meetings, passing resolutions that are clearly *ultra vires* and then seeking to enforce them . . . So one can't help but think that if there was a professional body, an organisation which you can campaign to and which holds itself out as having professional standards . . .

Conclusion

This chapter examines how similar issues in England and New Zealand concerning private residential neighbourhoods can arise from very different circumstances. Nether Edge was historically gated. Its conversion to a development was driven by one developer and required both major retrofitting as well as the construction of new housing. The result has been an exclusive development where the mix of old and new has created an attractive and coherent built environment. By contrast, Ambrico Place, with its mixed housing quality, lacks such coherence as, from the outset, there was no overall site development plan coordinating the work of the several developers. The resultant layout exhibits a blandness of style where the largely Mediterranean or American south-west influences do not sit well in the Antipodean landscape.

The research presented in this chapter has revealed that in both countries the planning systems that allow such neighbourhoods to be created require attention. In the English study the planning authority's lack of confidence in their own power to control the development meant that the national, regional and local policy aims of achieving mixed-tenure communities could not be achieved. In the New Zealand system, while there was consistency between local and regional policy, the development of the site revealed a policy vacuum which was, in part, a consequence of the effects-based approach that underpins local and regional planning. It was demonstrated that, in both cases, the councils responsible for granting planning permission for the developments to proceed lacked sufficient powers to impose conditions that would have resulted in better outcomes for the residents. In effect, the result was that the role of the councils became, almost by default, one of 'assisting' in the developer-driven process, rather than one whereby the developers and councils worked together in a partnership that took as its major focus the longer-term benefits and interests of residents.

A further issue demonstrated in the case studies is that of the inadequacies highlighted with respect to internal governance. The legal frameworks governing private residential neighbourhoods with shared ownership of common areas have proved to be advantageous to both developers and managing agents at the expense of residents. In both cases it is the developers who have the power to set up the conditions of internal governance. These are the arrangements that are in place when residents take up occupation. Despite the fact that residents have had no voice in the construction of these arrangements, such arrangements are extremely difficult to change and have a major impact on the long-term management of the developments. Even when the developers' connections have long since been severed, the initial legal arrangements, and the power relations embodied within them, continue. At this later stage, however, it is often the professional managing agents who, in a sense, have inherited the mantle of the developer, and who can continue to act in ways disadvantageous to residents.

Both case studies also revealed that residents were frequently unaware of the detail of the rights and responsibilities they had taken on by moving into these privately managed developments. The research highlighted numerous problems and much dissatisfaction on the part of residents with these internal governance arrangements. With respect to managing agents in England, the Commonhold and Leasehold Reform Act 2002 contains no provision for their regulation or licensing, despite a long history of complaints. They may be appointed as directors of the new commonhold associations. However, in New Zealand sufficient concern about the inadequacies of bodies corporate as a mechanism for managing private developments has resulted in a decision by central government to review the UTA 1972. It is likely that this review will contain provisions to regulate professional managing agents in some way.

Issues pertaining to private urban governance have, until now, been largely unrecognised and untested in the New Zealand context. However, in Auckland particularly, given increasing residential intensification, housing consumers are becoming more aware of the issues involved. In effect, a steep learning curve is in progress. It is surprising, given England's much longer experience of private residential developments, that many governance issues remain outstanding.

Bibliography

ACPO CPI Ltd (2003) *Secured by Design*. Online. Available HTTP: http://www.securedbydesign.com (accessed 12 February 2004).

Atkinson, R., Blandy, S., Flint, J. and Lister, D. (2004) *Gated Communities in England*, London: Office of the Deputy Prime Minister.

Auckland Regional Growth Forum (1999) *A Vision for Managing Growth in the Auckland Region: Auckland Regional Growth Strategy: 2050*, Auckland: Auckland Regional Council.

Blandy, S. and Robinson, D. (2001) 'Reforming leasehold: Discursive events and outcomes, 1984–2000', *Journal of Law and Society*, 28 (3): 384–408.

Department for the Environment, Transport and the Regions (1999) *An Analysis of Responses to 'Residential Leasehold Reform in England and Wales – Consultation Paper'*, London: DETR.

—— (2000) *Planning Policy Guidance Note 3, Housing*, London: HMSO.

—— (2002) *Improving the Standard of Residential Leasehold Management: Consultation*, London: DTLR.

Department of the Environment and Department of Transport (1992) *Design Bulletin 32: Residential Roads and Footpaths: Layout Considerations*, London: DoE and DoT.

Dixon, J. and Dupuis, A. (2003) 'Urban intensification in Auckland, New Zealand: A challenge for new urbanism', *Housing Studies*, 18 (3): 353–68.

Dixon, J., Dupuis, A., Lysnar, P., Spoonley, P. and LeHeron, R. (2001) *From Clay Pit to Community: A Study of Medium Density Housing in Ambrico Place, New Lynn*, report for the Waitakere City Council, Auckland: Massey University and The University of Auckland.

Dupuis, A., Dixon, J., Lysnar, P. and Mouat, C. (2002) *Bodies Corporate and Intensive Housing in Auckland: A Preliminary Assessment*, Report for the Auckland Regional Council, Auckland City Council, Manukau City Council, North Shore City Council, Rodney District Council and Waitakere City Council, Auckland: Massey University and The University of Auckland.

Government Office for Yorkshire and the Humber (2001) *Regional Planning Guidance for Yorkshire and The Humber*, Wakefield: The Yorkshire and Humber Assembly. Online. Available HTTP: http://www.goyh.gov.uk (accessed 12 February 2004).

Royal Institute of Chartered Surveyors (1997) *Code of Practice for Managers of Residential Accommodation*, London: RICS.

Sheffield City Council (1997) *Planning Brief for Nether Edge Hospital*, Sheffield: Directorate of Planning and Economic Development.

Waitakere City Council (1995) *Our Place, Our Future: Waitakere City Proposed District Plan 1995*, Waitakere, New Zealand: Waitakere City Council.

14

MORE GATES, LESS COMMUNITY?

Guarded housing in Russia

Sebastian Lentz

Abstract

In the mid-1990s gated communities first appeared on the Russian property market. Since then their number has been increasing constantly. Some wealthier members of society choose to retreat behind walls, fences and gates, protected by electronic devices, patrols and guard dogs. Such housing complexes are to be found in a growing number of large Russian cities, but the greatest number and concentration are in Moscow and its hinterland. The architecture and exterior design of these communities and their internal structure and organisation are extremely varied, so that it is not always easy to compare them with 'classical' gated communities in the West.

In a society that was permeated by the ideology of social equality and where the idea of 'community' was defined by values which were not bound to private property, the phenomenon is discussed in opposing ways: as a symptom of a sick, fragmented capitalist society or as a sign of a new freedom of choice, in this case to choose one's own neighbourhood.

Two types of interpretation are offered. According to the first, gated communities can be seen as a new phenomenon, related to rapid economic, legal and societal transformation. Prompted by real and imagined threats in a quickly changing, confusing society, they became a successful, new, private housing model. A process of innovation and diffusion is suggested, with Moscow as the initial node whence the phenomenon has been adopted by other cities and city-regions, depending on their success under the new economic order.

The second interpretation points out that gated communities are not a new phenomenon but were already established in Soviet society. In spite of official self-representation to the contrary and widespread perception in the West, Soviet society was deeply fragmented and spatially subdivided, the housing sector being a very good example. The leading cadres of the Communist Party, politics and the armed forces (*nomenklatura*) had access to elite housing, which included gated communities in the strict sense of the word, from city apartments in big housing blocks to elite dacha colonies. Thus the current rise of gated communities can be seen as an extension of a long-established trend, continued by the wealthier classes instead of political and functional elites.

Research on cultural patterns in transition

In social science research on post-Soviet society, two general types of interpretation are usually offered:

1 The empirical findings are declared to be new, an outcome of the changed general nature of society, and related to economic, social and political transition. This approach was typical during the first years of post-Soviet research, when the abundance of new trends and rapid changes were hard to grasp and the direction of development difficult to analyse. A lack of theoretical alternatives resulted in explanations that were mostly influenced by a polarised model of transformation, seeing the former socialist society and centrally planned economy as being on its way to some kind of Western modernity and capitalism (see Häußermann 1995). Confronted with the huge variety in processes of transition in eastern Europe, political scientists, economists and sociologists had to admit, however, that the new social and economic conditions could not be explained by a linear, polarised model of democracy and market economy versus centrally planned, authoritarian regimes.
2 The concept of modernising eastern European societies moving towards a 'Western model' had to be revised. With growing doubts about the explanatory value of this approach, historical dimensions were taken into account and new interpretations emphasised path-dependencies: decisions made in the past were recognised as structural factors in current evolution (see French 1995). Finally, the role of cultural patterns was introduced into the debate; they are seen as values, norms and actions learned by society in the past, being widely accepted and reproduced through the daily interaction of individuals and institutions. Lying below the surface of the structural process, they change very slowly.

For urban geography, dealing with the spatial reorganisation of everyday life in post-socialist cities, the 'cultural turn' offers an interpretative approach which looks beyond the immediate effects of transition, searching for long-term trends and the historical roots of current phenomena. For example, the extremely dynamic changes in city structure in Moscow and St Petersburg in the early 1990s were interpreted as adjustments to market forces. The new economic conditions, especially the introduction of the economic principle of ground rents into city planning and city development, were seen as triggers. Net rent gaps opened between the profit from existing uses and the potential profit from a new use, which led at first to rapid changes in the use of land and buildings, and then to new development and refurbishment. In most cases, the organisational change was accompanied by increased economic efficiency in land use, as is common in 'Western' cities with a free land market. However, many of the newly developing land-use patterns do not correspond to those of western European or North American cities. Structural reasons can only partially account for this, for example the major role played by public transport; housing, industry and commerce are still oriented towards these linear axes (Lentz 2002). Such arguments fail to account for the continuing extraordinary popularity of the centres of large cities for residential uses, from premodern times through the Soviet period and today. Traditionally, the superior provision of the city centre with urban amenities, especially in times of shortage, was accepted as an explanation for this pattern (Smith 1989). Thus the percentage of the population living in the city centre appeared surprisingly high to visitors from the West, even allowing for those who had no access to a new flat and so were forced to live in communal apartments in the city centre. With the introduction of the market economy, the supply of goods and amenities quickly improved, both at the periphery of the city centre and in the outer fringes of the city, so that these locations became more attractive to potential residents. The population of central Moscow decreased in the 1990s, when economically weaker residents were displaced by new land uses. These included business uses, especially high-ranking services and retailing, as well as the construction of luxury apartment buildings, for single adults, small households and families (Bater, Amelin and Degtyarev 1998; Gdaniec 2001; Lentz 2000, 2001). In the mid-1990s apartment rents in the centre of Moscow were almost as high as for office space (Rudolph and Lentz 1999: 37). In spite of high costs, the negative environmental implications of greatly increased private car traffic and the availability of attractive building sites in Moscow's immediate hinterland, demand for city-centre housing remains high. Is the persistence of this locational preference part of a cultural pattern where living in the centre has carried prestige for such a long time (see Bater 1984; Trushchenko 1995) that Muscovites are willing to accept very high prices despite growing access to private transport, which would otherwise be expected to stimulate suburbanisation? And is the spread of guarded housing over the last ten years not only fuelled by economic and social transformation but also founded on persistent cultural habits?

The urban geographical perspective on the phenomenon of 'guarded housing' is a specifically spatial one in that it represents either a new spatial pattern or a change in an existing one. The change in spatial constellation, according to one paradigm, results from social change in the broadest sense. In the context of the transformation of post-Soviet society, this phenomenological approach has the advantage of facilitating the rapid and reliable description of change in times when the sources usual in the social sciences (e.g. social statistics, income statistics) are unreliable or do not provide any insight into the trend being studied.

Several factors are, however, involved in the creation of the new spatial patterns, and these must first be identified. First, the providers (those who select the location for a new housing development and build it) are involved in the production of new spatial residential patterns. Second, those looking for housing contribute to the creation of demand by choosing a new home in a gated and/or guarded community. Thus both the privatisation of land and the newly developed private housing market (which allows the free choice of place of residence given adequate spending power) are responsible for the creation of new principles of social order and the dissolution or modification of existing value systems. This leads to uncertainty among the people with regard to their own social position, as new classification systems develop slowly and must be 'learned' or accepted. Such 'markers' of social order do not have to be created from scratch but, according to this hypothesis, stem from older cultural patterns underlying social value systems and influencing them. The cultural argument maintains that the spatial rootedness of a society is as important a mainstay as social order; i.e. spatial and social order are closely interconnected and this is reflected in social learning. Thus even in periods of radical change, a society will not necessarily 'invent' a totally new spatial order, but will orient itself as far as possible on existing patterns. Studies of 'guarded housing' in modern Russia must therefore examine the origins of this form of housing in socio-economic transformation, while also taking into account the historical and cultural models which facilitate their rapid acceptance in society.

The rise of gated communities in the private housing market

In the mid-1990s the first apartments in closed residential complexes came on the market in Moscow. Their numbers have risen steadily since and continue to rise. Likewise, the phenomenon emerged in St Petersburg at roughly the same time, but numbers increased more slowly than in the capital. Building types range from single apartment blocks or housing complexes with single-family houses to whole districts, and there are great divergences in the degree to which the communities are closed off. Empirical data was collected by the author for the period 2001 to 2003, including detailed inspection of guarded communities in Moscow and St Petersburg as well as interviews with architects, management personnel, investors and residents. The following examples from Moscow provide a typical overview but do not include all forms of scheme (see also Lentz and Lindner 2003).

Pokrovsky Hills

These terraced houses were built by a Texan investor who works on an international basis, and they correspond closely to the stereotypical image of a gated community. Security is provided by the surrounding wall and watchmen. The management, including maintenance and repair services, is exclusively English-speaking. The roughly 260 residential units were intended from the first for expatriate staff. With monthly rents of US$8,000 to US$11,000 for the houses (160–250 square metres), Pokrovsky Hills is one of the most expensive residential areas in Moscow. While the houses were initially only let to foreigners, now about 7 per cent of the tenants are Russian. One of the main attractions of the community is the Anglo-American school on the site; the investor offered the school part of the site for a new school building; hitherto it had been scattered over several less attractive locations in Moscow. Thus the children of Pokrovsky Hills no longer have a long journey to school. The success of the complex, which according to the management was completely filled from the start, can be partially explained by the sense of insecurity experienced by many foreigners in Moscow (Roth and Roth 2002: 93f, 98, 104). A shuttle service from Pokrovsky Hills to the nearest underground station helps to minimise contact with the outside world. The management tries to foster a sense of community among the inhabitants by organising a programme of activities, although the management itself is quite sceptical of its effectiveness in this respect.

Similar settlements of detached and semi-detached houses are now being erected in large numbers at a moderate distance from the city centre, in the north-western, western and southern sectors, which have traditionally had less industry and are thus less polluted residential areas. These communities are not especially designed for foreigners but aimed at local or mixed residents.

Donskoy Posad

Donskoy Posad ('suburb') is the name of an 18-storey building near the famous Donskoy monastery. It contains 84 apartments which are between 80 and 200 square metres in size as well as 2,500 square metres for commercial uses. Its infrastructure is not particularly remarkable; it consists of a small gym, a car wash and a laundry as well as a small, little-used playground. 'Safe housing close to the city centre' is the marketing pitch and shared activities among the residents are practically non-existent. Security is provided by video cameras and various access controls (some of them automatic), using magnetic cards, as well as a round-the-clock guard. The tenants consist of foreign businesspeople living alone in Moscow and childless couples.

Donskoy Posad is an example of the closed high-rise residential complexes which are typical in central areas of Moscow and other Russian cities. Several reasons seem to account for these. First, Moscow's City Development Plan is oriented towards high-rise development. Areas for detached and semi-detached housing are very rare and lie mostly at the edge of the city or outside its boundaries. Living there implies long commuting journeys, as the services sector, especially administration, has not yet been suburbanised to any notable extent and the road network is not able to cope with the dramatic increase in private motoring. Second, house construction in Moscow is still largely in the hands of the successors to the large construction combines. These are specialised in large-scale buildings. Finally, high-rise living is not associated with lower social status, as in many western European cities. Muscovites with high social status and high incomes find it acceptable to live in high-rise buildings, once they are well equipped.[1] This can almost certainly be explained by the fact that improvements in industrial housing construction in the Soviet Union were always implemented in high-rise buildings from the 1970s on (Gradow 1970; Baranov *et al.* 1976; Colton 1995: 487–91). Moving to such a building in the period from the 1970s to the 1990s meant an enormous improvement in comfort and living standards for millions of Muscovites. This is reflected in the fact that recently completed high-rise developments and those under construction were the most popular residential areas in Moscow (Colton 1995: 516; Veshninskii 1996: 8). The buoyancy of demand for guarded apartment schemes reflects the persisting positive image of high-rise housing in Russian urban culture.

Vorobyovy Gory

The DonStroy firm is erecting seven high-rise residential blocks with up to 35 storeys at Vorobyovy Gory ('Sparrow Hills'), a prestigious location near the Moscow State University and the *Mosfilm* studios on the Vorobyovy Gory. Underground car parks leave most of the site available to create attractive, landscaped grounds, based on an 'Alpine' landscape ideal; the advertising shows cedars, mountain pines, juniper, carnations and edelweiss, as well as the 'classic' park elements of flower beds and fountains. The services include typical elements of Soviet *mikroraion* planning such as all-day childcare facilities and a children's club; ice hockey, tennis and mini-football grounds; as well as indoor playgrounds and sports halls for teenagers. The complex also has innovative, indeed luxurious, services such as a computer centre, a fitness club, an aquapark with various health and beauty facilities, beauty salons and an indoor ski slope for children. Restaurants and a supermarket are also being built on the Vorobyovy Gory site.

The complex can only be entered by continually monitored gates with barriers. The outdoor areas, entrance areas, halls, underground car parks, stairs and lifts are monitored by video. As well as the guards there are concierges and other helpers who, for example, would carry shopping to a resident's front door. Outside, onsite workers are employed for day-to-day maintenance work in the complex. There is a car wash and a small-repairs garage for car-owners. As well as the usual cable and satellite channels, the management plans to have an in-house television channel which will document life in the complex.

Kvartal 75 and Kvartal 84/85

For about ten years a particular process in urban development has been triggered by the fact that the first generation of industrially mass-produced housing has exceeded the lifespan intended at the time of construction. Most of the residential areas built in the 1950s and 1960s are infamous for their substandard five-storey buildings (*pyatietazhki*). As they are at a middling distance from the city centre, and thus in zones where almost no building sites are otherwise available, they are particularly interesting to investors (Lentz 2002: 32). The old Kvartal 75 ('Urban District no. 75'), with a total area of 23 hectares, is being completely redeveloped over seven years. Only five or six of the total of

38 *khrushchoby* (a made-up word used by Russians, from Khrushchev and *trushchoby* – shack) will be retained. The project includes around 500,000 square metres of new development, 390,000 square metres of this being residential space in buildings with between five and 32 storeys. The plan is not to close off the district itself from the outside world, but to secure each individual new residential block. A decree made by Moscow City Council ensures that the current residents will get apartments in the new development. The infrastructure (shops, personal and household services, offices, a large business centre, childcare facilities, a youth centre, a cinema, a fitness centre and a swimming pool) will be financed by the proceeds from the sale of the apartments and business spaces. The open spaces between the buildings will be completely redesigned by the investor, who has a free hand in the provision of services and the landscaping of the open spaces. The Krost company is currently planning a second project similar in scale at Kvartal 84–85, which is to be completed by 2007 (Lentz and Lindner 2003: 56). The flats will be sold to private owners while a branch company of Krost, specialising in the management of buildings, will provide the services. It is normal that the residents' and owners' congregation evaluates the quality of the services and has the right to take over the management in its own administration or choose another company, if they are not satisfied with the actual performance. According to a site manager there have been several such cases in Moscow in recent years, so that there is already a concurrence between providers in this branch.

Summary of the phenomena

Expansion of guarded housing

A common feature of all the examples discussed above is that they are aimed at an upper middle- and upper-class clientele. In Moscow and St Petersburg, unlike the late 1990s, it is no longer a phenomenon solely affecting the elite (insofar as it is possible to apply this class concept to transforming Russian urban society – see for example Zaslavskaya 1996; Dittrich, Oswald and Voronkov 2002: 8–10). There is, however, a differentiation towards the 'top': the richest city residents have, as a rule, a house of their own in the hinterland, which like their spacious city apartment is guarded and secure. The fact is that new residential buildings in Moscow or St Petersburg are almost always closed or even guarded buildings, the exception being new public housing. The lowest common denominator of code-controlled doors and concierges is always adhered to, but is usually augmented by additional security measures and video monitoring.

Locational niches

The new housing complexes are located on the edge of the cities, or at a medium distance from the city centre in attractive areas, i.e. often in green belts which had not been built upon until the 1990s. These areas are often environmentally valuable open spaces, but their existence is not merely a product of corresponding intentions in urban development for improving the environment in the city centres. They developed as spaces between the *mikroraions* central to Soviet urban planning principles; the *mikroraions* in turn were developed in accordance with the linear structure of large-scale infrastructure (metro lines, main roads leading out of the cities, bus and tram lines, district heating, water pipes and so on). The new housing complexes with terraced houses, also oriented towards an upper middle-class clientele, can be built at a greater distance from the public transport routes, as their inhabitants are not dependent on easy access to public transport but often have a private car. In addition, the cost of new infrastructure can be transferred to the well-off buyers. From an urban planning point of view the development of these locations contributes to completing the development of the inner urban areas and the intensification of housing densities.

Socio-spatial differentiation

The new housing forms are to be found in areas of the cities where the residents' socio-economic status was already above average in the Soviet period. Such intra-urban differentiation should not have occurred, according to socialist planning ideology, but in fact there were already considerable differences in social structure and economic potential in the Soviet period (D.M. Smith 1989; G.E. Smith 1989; Vendina 1994: 14–15; Rudolph and Lentz 1999: 34–5). In Moscow, for example, the proportion of academics and highly qualified cadres was well above average in the south-western and north-western sectors of the city, while in the eastern half of the city the proportion of workers was higher, holding for other indices of social status as well (see Hamilton 1993; Vendina 1994: 38; Stadelbauer 1996: 114–16). Apart from the self-contained nature of the new complexes, their location reinforces segregation tendencies in the urban macrostructure (see figure 14.1 overleaf).

Freeing names from ideology

In the Soviet period, ideologically coloured or political-programmatic names, e.g. commemorating events of the revolution, were often used for urban districts and *mikroraions*, typical examples being Falcon, Red Guard, Komsomol. Nowadays most settlements and their names are emphatically apolitical. Old place names and field names are often used (see Vorobyovy Gory), as well as idyllic name such as Island of Fantasy. Occasionally, however, names such as Triumph Palace (a new high-rise residential complex in the north-west of the city) crop up, perhaps to reflect the residents' success and prosperity.

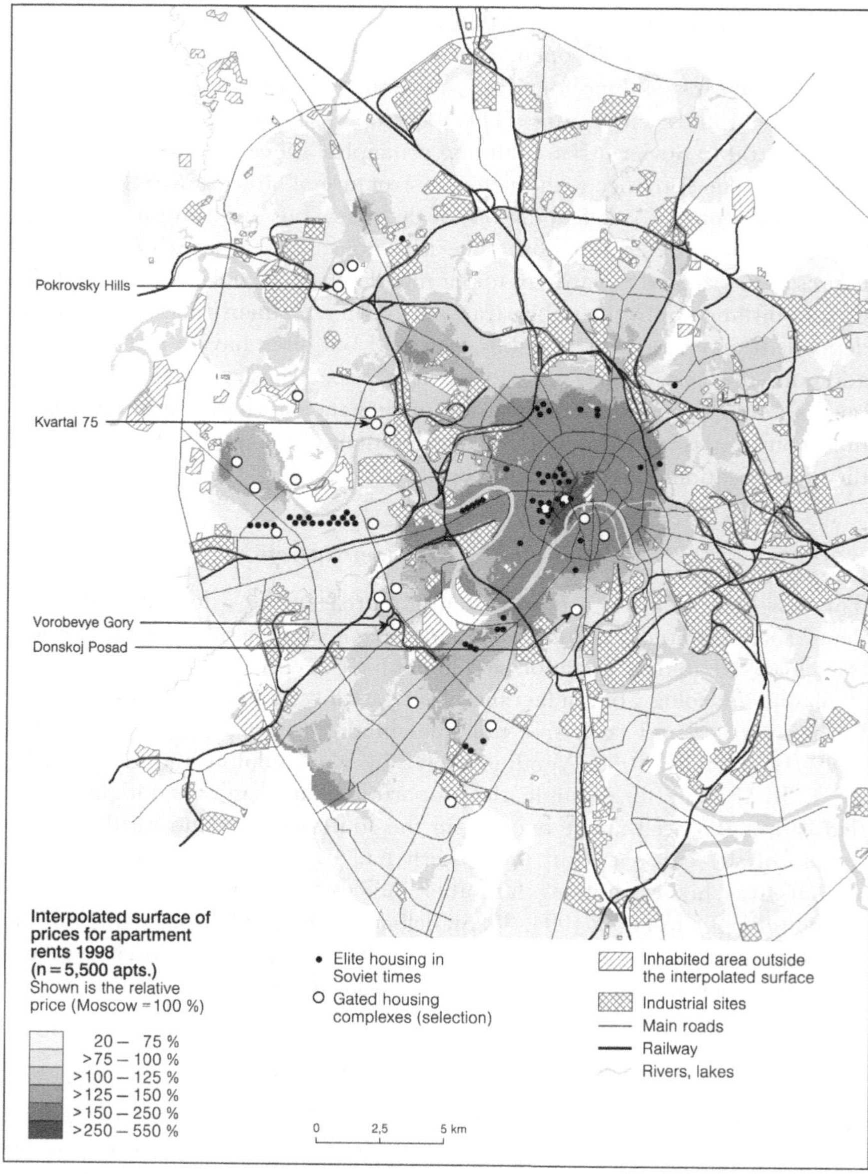

Figure 14.1 Congruence of the new locations in Moscow with old elite districts. Sources: the author; Colton 1995: 503

Interpretations[2]

It is frequently assumed that changes in the fabric of former socialist cities reflect the transformation of the social structure from a state of relative equality to a market economy structure with large social gaps in the 'capitalist' city.

And indeed, in the last 12 years, numerous phenomena have emerged in the post-socialist cities which clearly reflect social and economic splits in society, as for example the very varied reactions of the population to the new consumer worlds in the cities' centres (Argenbright 1999: 9–12).

As described above, new residential complexes in Moscow and St Petersburg now almost always provide some form of security monitoring. Furthermore, the complexes are provided with a range of services intended to ameliorate the problems of everyday life in a (post-socialist) city. It is almost impossible to buy a private apartment[3] in the country's two metropolises without monitoring and other services being an integral part of the rental or sale agreement. Thus the decision to move into such a complex does not per se demonstrate which service is most relevant to the new residents or whether the closed nature of the complex was the most important factor in this decision. Specific motives for moving into a building or complex are obscured by the services package. Rather, clients accept services which are of no particular importance to them; it is even possible that some inhabitants might accept disadvantages (in our case a monitoring of their own person around their flat) and still move into such housing, because there is no alternative to choose in what they define as decent housing.

In order to assess the social significance of the guarded-housing phenomenon more accurately, a detailed survey of residents would be necessary. For all kinds of obvious reasons it is difficult to carry out representative surveys. Managements who see it as their responsibility to protect their clients (residents) from everyday hassle do not usually permit direct, organised surveys to be carried out in their buildings. The providers are, however, more accessible. They are usually willing to discuss their housing concepts and their perception of the market. This can be supplemented by advertising brochures for the various complexes, from which one can draw conclusions as to a provider's competitive position in the market.

These indicate that the type of housing preferred by foreign economic elites could have provided the models for the first guarded housing complexes. It appears that the centre of innovation lies in the metropolis, where the largest number of foreign business representatives live. Talks with investors suggest a more far-reaching explanation: in the early phase of the private housing market little was known about Russian customers' wishes and requirements, so that investors to some extent planned the provision of services and so on based on what was known of foreign models. A representative from a firm operating such housing complexes emphasised the orientation towards American models: 'We don't have to reinvent the wheel'. Thus some investors may have designed their complexes without specifically knowing whether Russian buyers would appreciate such luxury. Standards then became established in local competition, and it is now impossible to ascertain the extent to which customers demanded the security aspect.

Given these research conditions, caution is advisable when considering the following interpretative suggestions on the security aspects of guarded housing.

'Closing off' or 'separation' may only be one among several decisive factors behind the growth of this new housing form.

In search of security I

The external appearance of many of the guarded housing complexes in Moscow and St Petersburg suggests that the desire for greater security was an obvious reason for their existence. Many of the complexes which were built in the mid-1990s were not only surrounded by high fences and walls, but also guarded by patrols, in some cases even watchtowers were built for the security guards. 'Security', however, implies more than just the protection of life and limb. It also includes the guarantee of the long-term functioning of 'normal' residential infrastructure, e.g. the provision of water, gas, electricity, heating, regular waste collection, lift maintenance and so on – services which frequently failed in formerly state-owned and now municipally owned housing in the period of transformation. This was intensified by the decline of traditional social values, manifest in increased vandalism, e.g. destruction of letter boxes and hallway lighting, which were only slowly replaced by the state buildings management companies (or not at all). Lastly, the effects of the private housing market must be taken into account; housing was formerly assigned by state authorities and usually kept for life, but now the residents of a block move and change frequently,[4] which makes it more difficult for neighbours to communicate norms and creates insecurity.

In the special situation of transformation, post-Soviet society was directly confronted with the 'losers' in social processes far more often than had hitherto been the case. Homeless people in particular, the so-called *bomzh*, who according to rumour moved into the cellars of large housing blocks, caused a feeling of insecurity. In the mid-1990s a series of bomb attacks on residential buildings shocked Russian city-dwellers. In Moscow in particular, the residents of many housing blocks began to organise their own security with watchmen, new security measures at doors and entrances, and concierge services. The phenomenon is not confined to large cities; in dacha settlements where large numbers of thefts and burglaries took place in the early 1990s, the settlers also often organised their own security services.

Ultimately it is of little significance whether a danger actually exists or is realistically assessed. The chief reason for the retreat behind gates and fences is a sense of threat. Foreigners are particularly susceptible because of the unfamiliar surroundings, especially when they are living and working in a place for a short time only, so that they do not become very familiar with the rules of everyday life beyond their immediate working environment. Thus a double circle of cause and effect comes into being: on the one hand closed housing complexes for foreigners are the product of their insecurity in a new environment, on the other hand their inhabitants isolate themselves from their surroundings to a marked degree, so that they take much longer to become familiar with them.

The weakness of city administrations

At the beginning of the 1990s the city administrations were virtually incapable of providing the basic infrastructure to make building land attractive to investors or to make the renovation of existing houses appear profitable. One solution was to transfer the responsibility for development to private investors. From their point of view, such investments could only be profitable if they developed large units, ranging from larger estates to whole city districts, a field in which they already had considerable experience. The developers passed the costs on to the prospective buyers, while guaranteeing them exclusivity by controlling entrance to the complexes. Seen like this, guarded housing is a sign of the weakness of public administration and its inability to organise the supply of housing neighbourhoods for the newly rich and the new middle class. This coincided with the abandoning of socialist planning ideals, which were replaced by the ideology of the market. Thus the state authorities' large-scale retreat from the design of city districts, which in Soviet times was a very important field of applied ideology, becomes more understandable.

In search of security II

A second possible interpretation of the security aspect relates to social norms and value systems. The transformation of Soviet society created an environment where familiar and acknowledged fixed points in society were rapidly invalidated. Thus functional-meritocratic criteria of social status were replaced by socio-economic criteria (Dittrich, Oswald and Voronkov 2002: 8f). According to this hypothesis, individuals and social groups try to compensate for the insecurity arising from this loss of orientation by creating formal order. Enclosed housing, where certain aspects of coexistence as well as the external order of the immediate surroundings are predetermined by comprehensive rules and regulations, gives a sense of security in a lifeworld that has descended into disorder. This sense of security is reinforced by the relative social and economic homogeneity of the residents in such complexes. At the same time, these complexes are no longer a state-directed recognition of services to society. Rather, in the manner of Foucaldian 'heterotopia' (Foucault 1997[5]) they become a means of expressing one's own 'otherness' (here in the context of new social rules, 'success') through one's place of residence and isolation from other sections of society.

This segregation creates an unusual tension in post-Soviet society, because Soviet society was ideologically egalitarian or based on the production collective (Shlapentokh 1989: 130). Collectivism implied the dedication of personal energy to the collective, where the collective served as the moral point of reference. Accordingly, the aim of town planning was to prevent segregation arising in the first place, but it unintentionally encouraged it by building workers' housing close to their place of employment. This was intended to keep journeys to work to a minimum and to extend the sphere of influence of the workers' collectives to communal leisure activities. The persistent spatial isolation of elites in the period of transformation reinforces ideologically coloured prejudices, especially among those members of the intelligentsia who see socio-spatial segregation as a characteristic of pathological capitalist systems.

Cultural patterns I

The criticism of the separation of the elites from 'common' society can be found in the more intellectual media, but it does not find a very enthusiastic response in broad sectors of post-Soviet society. Such general criticism as does exist is directed towards the lifestyle of the 'New Russians' who make a great display of their wealth. But this criticism is associated more with the source of this wealth than with segregation (Neidhart 2000). It must, however, be borne in mind that this criticism has also contributed to the sense of insecurity among the new upper and middle classes. But why is the criticism of segregated housing not shared by wider sections of society? One possible reason is that the new freedom of individual choice is welcomed and accepted by a majority. New concepts and values of privacy can thus be assumed to be established among the population. A more sceptical interpretation is based on the remarkable reduction in socio-political awareness and active social involvement in the period of transformation, giving way to an attitude that tolerates individualistic values and corresponding behaviour. However, numerous studies from as far back as the 1970s, e.g. Bater (1977), Zaslavsky (1982), Shlapentokh (1989) and Lewada (1992), indicate that this is not a new phenomenon; an increasing retreat of individuals from the public–political sphere of Soviet society could already be observed in the 1970s, especially involving a reorientation from the collective to the family as the central unit of life after work. The physical or spatial preconditions for this attitude are to be found in the mass produced housing, which from the mid-1960s on gave millions of Soviet citizens an apartment of their own which they did not have to share with others (Baranov and Beloussov 1976).

The existence of self-contained housing forms in the Soviet Union reflects even older historic patterns. Contrary to a widespread perception of the Soviet system, space there was largely divided into compartments, access to which was strictly regulated and monitored. This was true of housing. Shortly after the Revolution, the housing of the functional elites was segregated. This principle was initially only applied to the secret service, but was later extended to many other professional groups. Political and party elites also retreated to housing complexes that fulfilled the classical criteria of gated communities (and ghettos). They had a wide range of their own facilities, educational services and leisure features, as well as guarded housing. Thus the hypothesis emerges that these and similar housing forms are widely socially accepted today because they had precursors which made segregated, guarded, secure housing appear desirable in spite of egalitarian tendencies. For Soviet citizens, living in closed or guarded complexes was not a sign of social ostracism but rather showed that the inhabitants were special in a positive sense (see Schlögel 2003).

They either had a higher social status, such as the members of the *nomenklatura*, or were of outstanding merit, such as artists or top sportsmen and sportswomen. Another group were the foreigners, who lived in segregated blocks guarded by the militia. They were materially better off than Soviet citizens and enjoyed a lifestyle and goods not generally available, including foreign travel and other important prestigious features of consumption. Thus the acceptance of segregated, guarded housing can be interpreted as an example of the persistence of long-established values.

Cultural patterns II

A tentative concluding interpretation, bearing in mind the tentative empirical base, is that guarded housing not only means secure housing for its inhabitants, but also implies a monitoring of residents' own behaviour. In Western societies a certain degree of scepticism apparently exists with regard to the monitoring of individual lifestyles. The residents of guarded housing complexes in Russia today apparently do not object to their own everyday comings and goings being registered and monitored by guards and cameras; unless of course other motives for choosing this type of housing outweigh possible reservations about the permanent surveillance. One can only speculate whether accustomisation to the supervision of everyday life by *dezhurnye* ('concierges') over long periods, and the very system of *dezhurnye*, also play a role here.

Notes

1 This opinion was voiced by several interviewees on the provider side, e.g. by an architect from the construction company Konti on 26 March 2002.
2 The interviews on which the interpretation is based were conducted in St Petersburg by Konstantin E. Aksyonov (University of St Petersburg), whom I would also like to thank for long discussions on the reasons for the growth of guarded housing in post-socialist St Petersburg. I would like to thank Olga I. Vendina (Russian Academy of Sciences, Moscow) and Peter Lindner (Yale/Erlangen) for their cooperation in Moscow.
3 Rental occupation is rare in this segment of the market, except in the case of foreigners.
4 Nevertheless, the frequency of moves in Russian cities is still well below that in western European or North American cities (Struyk and Romanik 1995; Struyk and Kosareva 1997; Lee and Struyk 1996), as much of the existing housing became the property of the tenants, who continue to use the apartments themselves.
5 For a detailed examination of gated communities as heterotopia see Hook and Vdroljak 2002.

Bibliography

Argenbright, R. (1999) 'Remaking Moscow: New places, new selves', *Geographical Review*, 89 (1): 1–22.

Baranov, N.V. and Beloussov, N.V. (1976) 'Edilizia urbana ad uso abitativo', in N.V. Baranov, N.V. Beloussov, C. Cesari, G. Cuzzani, G. Gresleri, G. Lombardini, P. Maggi, C. Monti, S. Panzarasa, C. Pea, M. Petrignani, V. Quilici, G. Ronzani, G. Trebbi, G. Turcini and M. Zaffagnini: 'Esperienze ed orientamenti dell'edilizia abitativa sovietica', Mostra organizzata dall'Ente Autonomo per le Fiere di Bologna . . . nell'ambito del 12. SAIE, Salone Internazionale dell'Industrializzazione Edilizia, E.A. Fiere di Bologna, Bologna, pp. 13–21.

Baranov, N.V., Beloussov, N.V., Cesari, C., Cuzzani, G., Gresleri, G., Lombardini, G., Maggi, P., Monti, C., Panzarasa, S., Pea, C., Petrignani, M., Quilici, V., Ronzani, G., Trebbi, G., Turcini, G. and Zaffagnini, M. (1976) 'Esperienze ed orientamenti dell'edilizia abitativa sovietica', Mostra organizzata dall'Ente Autonomo per le Fiere di Bologna . . . nell'ambito del 12. SAIE, Salone Internazionale dell'Industrializzazione Edilizia, E.A. Fiere di Bologna, Bologna.

Bater, J.H. (1977) 'Soviet town planning: Theory and practice in the 1970s', *Progress in Human Geography*, 1: 177–207.

—— (1984) 'The Soviet city: Continuity and change in privilege and place', in J.A. Agnew, J. Mercer and D.E. Sopher (eds) *The City in Cultural Context*, Boston, MA: Allen and Unwin: 134–62.

Bater, J.H., Amelin, V.N. and Degtyarev, A.A. (1998) 'Market reform and the central city: Moscow revisited', *Post-Soviet Geography and Economics*, 38: 1–18.

Colton, T.J. (1995) *Moscow: Governing the Socialist Metropolis*, Cambridge, MA: Belknap Press of Harvard University Press (Russian Research Center Studies 88).

Dittrich, E., Oswald, I. and Voronkov, V. (2002) 'Einleitung: Wandel alltäglicher Lebensführung im postsowjetischen Russland', in I. Oswald, E. Dittrich, and V. Voronkov (eds) *Wandel alltäglicher Lebensführung in Russland. Besichtigungen des ersten Transformationsjahrzehnts in St. Petersburg*, Münster, Hamburg, London: Lit-Verlag (Osteuropa: Geschichte, Wirtschaft, Politik 31): 7–23.

Foucault, M. (1997) 'Of other spaces: Utopias and heterotopias', in N. Leach (ed.) *Rethinking Architecture: A Reader in Cultural Theory*, London: Routledge: 350–5.

French, R.A. (1995) *Plans, Pragmatism and People: Legacy of Soviet Planning for Today's Cities*, London: UCL Press (Changing Eastern Europe 2).

Gdaniec, C. (2001) 'Ostozhenka – a Moscow district in transition: Reflections of post-Soviet restructuring processes', *Ethnologia Europaea*, 31 (2): 41–58.

Gradow, G.A. (1970) *Stadt und Lebensweise*, Berlin: VEB Verlag für Bauwesen.

Hamilton, E.L. (1993) 'Social areas under state socialism: The case of Moscow', in S.G. Solomon (ed.) *Beyond Sovietology. Essays in Politics and History*, Armonk, New York: Sharpe: 192–225.

Häußermann, H. (1995) 'Von der "sozialistischen" zur "kapitalistischen" Stadt', *Aus Politik und Zeitgeschichte*, B 12/95 (17 March): 3–15.

Hook, D. and Vrdoljak, M. (2002) 'Gated communities, heterotopia and a "rights" of privilege: A "heterotopology" of the South African security park', *Geoforum*, 33: 195–219.

Lee, L. and Struyk, R.J. (1996) 'Residential mobility in Moscow during the transition', *International Journal of Urban and Regional Research*, 20: 656–70.

Lentz, S. (2000) 'Die Transformation des Stadtzentrums von Moskau', *Geographische Rundschau*, 7/8: 11–18.

—— (2001) 'Zur kulturgeographischen Interpretation aktueller Stadtentwicklungsprozesse in Moskau', in H.-H. Höhmann (ed.) *Wirtschaft und Kultur im Transformations-*

prozeß. Wirkungen, Interdependenzen, Konflikte, Bremen: Edition Temmen: 239–55 (Analysen zur Kultur und Gesellschaft im östlichen Europa 11).

—— (2002) 'Moskau – aktuelle Stadtstrukturentwicklungsprozesse', *Geographie und Schule*, 24 (136): 29–34.

Lentz, S. and Lindner, P. (2003) 'Die Privatisierung des öffentlichen Raumes – Soziale Segregation und geschlossene Wohnviertel in Moskau', *Geographische Rundschau*, 55 (12): 50–7.

Lewada, J. (1992) *Die Sowjetmenschen 1989–1991. Soziogramm eines Zerfalls*, Berlin: Argon.

Neidhart, Ch. (2000) 'Russlands neue Räume', *Neue Züricher Zeitung*, 246 (22 October): 63–5.

Rudolph, R. and Lentz, S. (1999) 'Segregationstendenzen in russischen Großstädten: Die Entwicklung elitärer Wohnformen in St. Petersburg und Moskau', *Europa Regional*, 7 (2): 27–40.

Roth, J. and Roth, K. (2002) 'Leben in Moskau: Deutsche Entsandte und die deutsche Gemeinde in der russischen Hauptstadt', in A. Moosmüller (ed.) *Interkulturelle Kommunikation in der Diaspora. Die kulturelle Gestaltung von Lebens- und Arbeitswelten in der Fremde*, Münster, New York, München, Berlin: Waxmann: 77–111.

Schlögel, K. (2003) 'Der Mercedes-Stern auf dem "Haus an der Moskwa". Vom Kommunehaus zur bewachten Wohneinheit', *Frankfurter Rundschau*, 21/2003 (25 January): 41.

Shlapentokh, V. (1989) *Public and Private Life of the Soviet People: Changing Values in Post-Stalin Russia*, New York, Oxford: Oxford University Press.

Smith, D.M. (1989) *Urban Inequality under Socialism: Case Studies from Eastern Europe and the Soviet Union*, Cambridge: Cambridge University Press.

Smith, G.E. (1989) 'Privilege and Place in Soviet Society', in D. Gregory and R. Walford (eds) *Horizons in Geography*, Basingstoke, London: Macmillan Press: 321–40.

Stadelbauer, J. (1996) 'Moskau – Post-sozialistische Megastadt im Transformationsprozeß', *Geographische Rundschau*, 48 (2): 113–19.

Struyk, R.J. and Kosareva, N. (1997) 'Is the market working?', in R.J. Struyk (ed.) *Restructuring Russia's Housing Sector 1991–97*, Washington, DC: The Urban Institute: 127–32.

Struyk, R.J. and Romanik, C. (1995) 'Residential mobility in selected Russian cities: An assessment of survey results', *Post-Soviet Geography*, 36: 58–66.

Trushchenko, O.E. (1995) *Prestizh Tsentra. Gorodskaya sotsial'naya segregatsiya v Moskve*, Moskva: Izdatel'stvo (Socio Logos).

Vendina, O.I. (1994) *Moskau: Eine Stadt verändert ihr Gesicht*, Köln: Bundesinstitut für ostwissenschaftliche und internationale Studien (Berichte des Bundesinstituts für ostwissenschaftliche und internationale Studien 43).

Veshninskii, J.G. (1996) '"Luchshii" – Belorusskii, "chudshii" – Kazan'skii. Tak bylo i est", *Segodnja*, 95 (1 June): 8.

Zaslavskaya, T. (1996) 'Transformatsiya sotsial'noi struktury rossijskogo obshchestva', in T. Zaslavskaya (ed.) *Kuda idyot Rossiya? Sotsial'naja transformatsiya postsovetskogo prostranstva*, Moskva: Aspekt Press.

Zaslavsky, V. (1982) *In geschlossener Gesellschaft. Gleichgewicht und Widerspruch im sowjetischen Alltag*, Berlin: Wagenbach.

15

CONCLUSION

Dynamic urban order and the rise of residential clubs

Chris Webster and Georg Glasze

The first conclusion we wish to draw from this collection of papers is that private neighbourhoods are beyond doubt a highly significant feature of post-modern urbanism. They are a global phenomenon, they signify a distinct shift in the way cities are organised, they represent an extreme decentralisation of decision-making, they create new socio-spatial divides, their popularity has spawned a commodified neighbourhood industry, they create micro-territories with their own local constitutions, they represent a new model of collective action in the coordination of demand for public goods, they break down traditional social geographies and create new ones, they provide an alternative model for financing civic goods, they do for home buyers what the package-tour industry has done for individual holiday buyers, they enhance the conspicuousness of home-based consumption, they alter patterns of urban insecurities and they create new micro-societies.

For all these reasons and more, they should be taken seriously and studied for their potential impacts on cities and societies. The authors in this book have chosen to emphasise these themes and issues in various ways. Some are pessimistic in their overall assessment of the effects of gating; others are more optimistic. One of our purposes in compiling a range of commentaries from writers taking contrasting perspectives is to challenge readers to reflect on the differences in approach and, in particular, to distinguish between normative and positive arguments; private and social cost considerations; behavioural, action-oriented, discourse-oriented and structural analysis; deterministic and methodologically individualistic institutional analysis; local and global explanations and prognosis; and short- and long-term analysis. In the following section we select some of the key themes in the book and reflect on the importance of time in the analysis of urban change. Then we consider the configuration of local and global factors that have contributed to the rise of the residential club as a popular institution of urban governance. Finally, we conclude with some reflections on the complexity and entropic nature of urban evolution.

Stasis or dynamism?

Virginia Postrel (1999) has suggested that the old left–right ideological divide is being replaced by a division based on a preference for stasis versus dynamism. Some people are more comfortable with change than others. This also amounts to saying that people take positions on the basis of what time frame to use in assessing social costs and the need for government intervention. This cuts several ways. Sustainability as an ideology demands an intergenerational time frame and urges restraint on present consumption for the sake of future consumers. On the other hand, threats to present order are often considered within short time frames that give inadequate consideration to the ability of social systems to adapt.

Evan McKenzie's analysis of US common-interest developments (CIDs) illustrates well the importance of the time frame (McKenzie 2003). It stands out among the more pessimistic commentaries because of its focus on the current operational problems of contractual communities and the problem of spiralling private, as opposed to social, costs. But its context is historical and dynamic in the sense that it conveys something of how institutions of the state and the private neighbourhood co-evolve. From what McKenzie has so far observed, he is not convinced that public and private institutions will be able to control adequately and allocate fairly the costs of running private cities. He notes that state laws are emerging which, for example, attempt to keep CID litigation out of the public courts and require developers and homeowners' associations (HOAs) to reserve adequate reinvestment funds. But privatopia through McKenzie's lens still looks like a costly model of organising urban society. One implication of his argument is that the phenomena may be short-lived. Once individuals discover just how costly it is in real terms (taking into account the probability of having to take or fight legal action, including eviction proceedings, and the likelihood that the costs of CID litigation will become increasingly internalised within the sector and therefore passed on to CID residents) then CID popularity may wane. More than this, where future private costs are not adequately planned for, private neighbourhoods may implode with bale-out costs to the wider public. The crucial variables in this analysis include the degree to which the costs of HOA litigation and reinvestment can be internalised within the industry and the degree to which the state needs to regulate. The latter includes the regulation of municipal governments by higher authorities to mitigate the dangers of municipal-government opportunism. The time frame is crucial to any analysis of the institutions governing the private-neighbourhood industry. Over time, McKenzie's commentaries have shown us that things are getting both better and worse. The number of horror stories about HOA governance and private regulations grows with the number of CIDs – leading to a critique of HOAs as 'shareholder democracies' (Glasze 2003b). On the other hand there is a better understanding of the failures of the private governance market and a growing raft of laws to address some of these. Detailed accounts of the problems that eventually give rise to new state laws, and of those that for one reason or another do not, are essential. What really matters in the long run is the degree of institutional responsiveness, the institutional gaps that may emerge, the reasons for and consequences of those gaps and the costs of supporting these institutions. It is possible, for example, that the legal infrastructure needed to support private urban government may be as cumbersome and costly to society (or more so) than that needed to support public government. In the end the main advantage offered may be greater choice of local government – for those who can afford it. But as the industry matures and explores ever more sophisticated products it may be that private cities grow in size (including by merger, partnership and federation of non-contiguous communities) and that choice and competition diminish. What we would then be left with is something that looks very much like conventional public government.

Fred Foldvary's optimism about such matters rests on his libertarian ideology. Voluntary government over urban affairs is, for him, better than coercive government. The high private cost of private governance is a matter for individuals to assess and markets to sort out over time. He would want to compare these costs with the hidden costs of publicly governed neighbourhoods – which are largely unknowable. Herein lies an interesting challenge to the position Foldvary represents. Faced with the full costs of managing their own affairs, including the costs of private mediation, private courts and infrastructure renewal, residents may prefer to offload some of these responsibilities back to the state. Given the freedom to choose between private and publicly governed neighbourhoods, homeowners may choose to free-ride, or to cherry-pick liabilities, keeping those that are within their means and offloading those that are not. This is a prognosis consistent with the political economy of public-choice writers and sits a little awkwardly with a strongly libertarian position. The same tension is there in the analysis of the social costs of private neighbourhoods that so many writers make so much of. Individuals may be expected to prefer forms of socio-spatial organisation that minimise the balance of net social costs they themselves bear. The markets in gated communities around the world have arisen with little acknowledgment of any social costs that may be imposed on other communities such as displaced crime and displaced traffic (interestingly, the issue of displaced tax spend should in principle be a positive spillover effect – unless, of course private communities are so successful at lobbying for tax exemptions or reductions, or at cherry-picking liabilities, that the net value of tax-funded urban services in publicly governed areas goes down). The welcome given to consensual local government by libertarians and indeed by the millions of homeowners now living in private neighbourhoods must be tempered, therefore, by an honest appraisal of the net social costs and benefits borne by wider society. It is true that equally consensual means of tackling some of these externality problems may emerge. Walls and gates are a consensual (market) solution to supplying security to a group of well-off Santiago residents, for example, but the development of a rich community abutting a poor community creates positive externalities of employment and service opportunities. Such locational externalities can be captured by individuals through private (voluntary) labour contracts (poor people working as maids and gardeners) and through loyal patronage (rich people patronising the services supplied in the poor neighbourhood) (Salcedo and Torres 2004). Contractual solutions to negative externalities caused by gated communities are less easy to find or imagine. This is so because of time and space dimensions. The problems of displacement, fragmentation and disruption are typically alleged to be longer-term social costs and the only agency that could negotiate on behalf of those affected would be government. Governments, however, represent present, not future, constituents. On the other hand, those who would suffer from the costs of any displaced urban problems are dispersed and are individually unlikely to perceive any marginal costs to them as being significant. The externalities caused or likely to be caused by gating should not therefore be sidestepped just because there is no immediate public outcry from the rest of the city. A spillover effect from a city's enclosed neighbourhoods is, like any other externality, an issue of missing markets (Arrow 1951, 1970). A resource valued differently by different groups but not governed by clear ownership rules cannot be efficiently traded or negotiated and gives rise to costly competition (Webster and Lai 2004). That competition is worked out by competing social groups over time and only over time can the nature of the problem and the value and fairness of alternative solutions be judged – often with a degree of private and public experimentation. An urban externality problem is likely to prevail until such time as it becomes sufficiently costly to some group or another for rules to emerge to address it. When that happens, the rules will never be acceptable to all parties (Arrow's impossibility theorem) but will reflect dominant interests and prevailing decision-making processes.

The evolutionary perspective in which institutions arise in response to externalities is not necessarily a fatalistic one. There is scope for analysis of the processes involved in order to inform public and private responses including market responses and policy at different levels of government.

An aspect of this dynamic that is of particular pertinence to the current debate is the idea that the institutional gaps that may grow while a social response is awaited may set in motion inferior patterns of socio-spatial development that are hard to reverse. Here is a Gordian knot that seems to have to be cut on the basis of ideological preferences and presupposition. Do we take present action to avert an imagined future problem even though we know neither the likelihood nor the size of the problem, nor the manner in which urban society will adapt to the problem if it is left to manifest more fully? Societies with liberal democratic constitutions and cultures that elevate the rights of individuals are likely to give private neighbourhoods the benefit of the doubt. Societies with democratic traditions which emphasise the common welfare, it may be assumed, may play safe and seek to limit the growth of a socio-spatial pattern of resource allocation that risks longer-term harm to society as a whole. In societies with missing or very deficient institutions assuring democratic decision-making it is likely that the emergence of private neighbourhoods serves the interests of those with the power to influence their emergence. This includes both the alliances of political and business interests that gain from the land development process and individual investors and home-buyers who gain use or investment value from buying into such schemes. Many critiques of gated communities view them as shifting urban resources in favour of more powerful groups (represented in our volume, for example, by Wehrhahn and Raposo's study of Lisbon and Madrid and Giroir's chapter on Beijing). This kind of analysis is valuable for describing new socio-spatial patterns of consumption and inequalities. In particular, several authors have made a link between the rise of private cities and the crisis of Fordist economics, the global economic restructuring process and the resulting shift in socio-spatial class structure. As a consequence of economic restructuring, labour categories and incomes become increasingly polarised, with a growing underclass on the one hand and a highly qualified transnational elite on the other. In urban debates this has become the dual-city hypothesis. It is not dissimilar to other dualist analyses that have focused on changing group boundaries and power balances, for example between urban and rural, inner and outer city, black and white, and male and female. Viewed as evidence of change in the patterns of wider social structures, private neighbourhoods are characterised as 'citadels' (Friedman and Goetz 1982) and 'touching-down places' (Marcuse 1997) of the transnational elite. Dear and Flusty (1998), Soja (2000) and others have extended the idea to that of a fragmented city. Los Angeles, for example, is said to illustrate how a deregulated and flexible economy and a pluralisation of lifestyles leads to the development of a 'postmodern urbanism' made up of enclaves, which are socially and functionally largely autonomous. In 1998 Dear and Flusty published a model that visualised these new urban spaces. The 'gated communities' are presented as one enclave alongside spaces like theme parks, shopping centres or ethnically defined quarters.

There is a sense in which this style of analysis is dynamic but insufficiently so. It presents a comparative static view of society in which a new state is compared unfavourably with an old state in a limited set of dimensions. It is non-dynamic in a number of important respects. One is the feedback effects that interlink international and domestic housing markets. Another is its failure to consider the possibility that private neighbourhoods may in some situations be a locally emergent phenomenon, explained more powerfully by local cultural values and the inadequacies of extant institutions. A third and related respect is the lack of consideration given to local institutional evolution. Urban policy, government regulations and the laws that govern private transactions are never static. Neither are the organisational structures that deliver them. They respond to efficiency and equity problems of resource allocation as defined by specific groups in specific cultural and constitutional contexts. An analysis that views gated compounds merely, or only, as enclaves of the rich is institutionally static. Domestic housing markets will adapt to the presence of high-end estates and if, as seems to be the case, the genre appeal crosses income divides, entrepreneurs will surely find a way of providing neighbourhood clubs to other income groups. Studies in the US (Frantz 2000) and elsewhere (Kanitschneider 2002; Kohler 2002) have shown that it is not only the elite who are moving into gated housing estates but that the trend is followed also by many households of the medium income range. At least in societies with a certain degree of democratic decision-making, municipal, state and national governments are unlikely to sit back and watch private neighbourhoods impose unacceptable costs on wider society. The interface between private institutions within contractual neighbourhoods and public institutions will co-evolve in complex and unpredictable ways (see for example Chen and Webster 2005).

The argument of a global economic restructuring leading to new fortified elite spaces stands in the tradition of neo-Marxist political economy. This macro-theoretical approach may offer an explanation of guarded housing estates targeted at the international market and may well partially explain why private neighbourhoods re-emerged so strongly at the end of the 20th century. However, empirical studies have shown that the dualisation of societies might not be the most important factor for the development of gated housing estates. The development of income distributions in some of the so-called global cities is not consistent with the idea of the dual city (Hamnett 1994). There are many regional variations in the style and extent of private neighbourhoods that are not adequately explained by this thesis. This includes the sparseness of such developments in some global cities like Tokyo or Paris and the presence in many cities of a strong contemporary domestic market, with historical antecedents, catering for middle- and low-income residents (Yap and Sakchai 2000; Webster, Wu and Zhao in this volume). To understand these more subtle local causes and consequences of private cities a stronger model of individual and state behaviour is required; one that permits an institutional analysis of both governments and markets over time as for example presented by Glasze's chapter on the relation between confessionally segmented patterns of social interaction in Lebanon and the rise of guarded residential complexes.

The final theme we wish to reflect upon is the discourse of fear, emphasised in this volume particularly in the chapters by Setha Low and by Ulrich Jürgens and Karina Landman. Jürgens' and Landman's analysis in this volume and elsewhere is a considered response to well-evidenced growing insecurities in South Africa. Low's work is based on conversations within gated communities and captures important subjective evidence that contributes to a dynamic understanding of the causes and consequences of gating. However, in as much as analysis which focuses on 'insecurity' laments sometimes the loss of some previous era of peace and harmony and stresses the divisive potential of legal neighbourhood boundaries, then it shares something in common with the dual-city hypothesis. It risks presenting a comparative static analysis that misses vital processes of social adjustment and evolution. For us it seems very important that research on the relations between insecurity and private neighbourhoods clarifies the distinction between objective risks and subjective insecurities. Research in critical criminology and critical criminal geography has shown that many societies show a trend towards growing (perceived) insecurities while statistics on crime are stagnant (see for example the contributions in Glasze, Pütz and Rolfes 2005). In many cities around the world there is a growing discourse on insecurity which shapes individual and social perception of risks. This discourse on urban insecurities seems to fuel the growth of private neighbourhoods and at the same time seems to be fuelled by this radical transformation of urban space. Therefore research on private neighbourhoods should seek to avoid simplistic analysis which uncritically ascribes a rise in security-related infrastructures and services like gates and guards to a rise in objective risks. Empirical work on the social costs of compounded fear and security is as important to informed debate as the empirical work on private costs represented by McKenzie in this volume and elsewhere (see also Briffault 1999; Walters and Kent 2000; Deng 2002, 2003a, 2003b). If private neighbourhoods cause irretrievable social fragmentation and a rise in the fear of others, then we need to know about it. Similarly, if the net effect of privately organised security is less total security in society then this should be evidenced and the story told. As with other alleged social costs, however, the questions are subtle and need to be guided by well-formed theory. We would like to know, for example, how the discourse on urban insecurities differs from one region to another and how this influences the acceptance of and demand for secured neighbourhoods. Several authors have asked if children socialised in private and socially homogeneous neighbourhoods might develop a greater fear of others than children growing up in 'normal' neighbourhoods – here, long-term interdisciplinary research projects seem to be necessary. The fundamental question is a comparative one: does territorial enclosure fuel fear of others through a dynamic qualitatively different to similar processes working in conventional neighbourhoods and cities?

As private neighbourhoods change patterns of segregation another question arises: do the functional ties that can develop between proximate neighbourhoods, especially with different income levels, mitigate rising fear? Given that nobody wants to live with fear, do practices and institutions emerge naturally within cities of residential 'clubs' that seek to address this negative by-product? The study by Salcedo and Torres (2004) is interesting in this respect. It is ethnographic, like Low's, but finds that residents in two neighbouring communities, one poor, one rich, have positive attitudes towards people in the other community based on a sense of mutual dependency. Here is a suggestion that spatial scale of segregation and functional social ties may matter more than walls and gates.

Clubbing together in cities – local and global explanations

Local gated housing markets have to be understood in a dynamic framework and this means understanding the changing local institutional contexts that give rise to and shape the institution of the residential 'club'. The economist

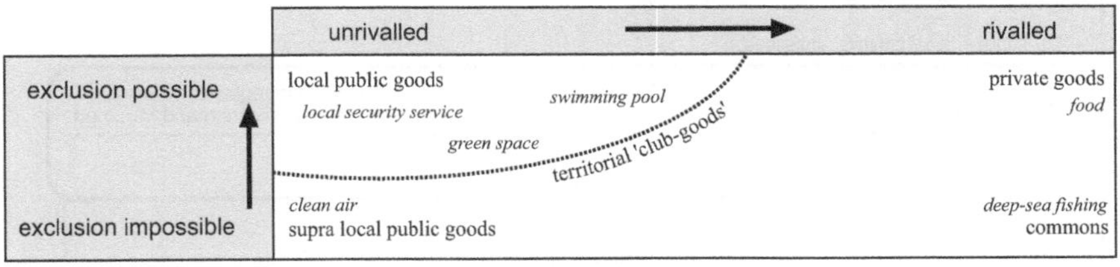

Figure 15.1 Private goods, public goods and 'club goods'. Source: Glasze 2003a

Charles Tiebout has shown in his seminal 1956 paper that some of the collective goods generally referred to as 'public' are better described as local public goods. Citizens (and neighbourhood residents) form spatially defined markets and consumption of shared resources falls off with distance. Green spaces, local security patrols or public swimming pools effectively serve only those people located near enough to enjoy them without incurring costs of travel that outweigh the benefits of consumption. The economic function of a private neighbourhood is to supply local public goods in quantities and mixes, and at a quality, that should match the preferences of residents. A gated community or a condominium apartment complex is a residential club with a territorial boundary. The economic theory of clubs was first formalised by James Buchanan (1965) who was interested in generalising Paul Samuelson's economic theory of public goods (Samuelson 1954, 1955). The latter made unrealistic assumptions about the degree of 'publicness' of most co-consumed goods and Buchanan's work established the idea that shared goods could be supplied efficiently so long as the number of consumers, access to the good and the quantity of it could be controlled (figure 15.1).

The entrepreneurial club, with rules governing membership (exclusion/inclusion), the payment of fees and collective behaviour of members, is an institution capable of supplying shared goods more efficiently than the institutions either of private commodity markets or of government. What Buchanan demonstrated theoretically the real-estate industry and home-buyers were quick to discover in practice, where and when the conditions were right for these markets to develop. The legal right to exclude non-contributors and the obligation to pay fees on a regular basis solves the problem of free-riding. Local decision-making should in principle make the continued supply of local public goods responsive to the preferences of residents and manage congestion more accurately. The manner in which neighbourhood goods are supplied and consumed and the physical and to some extent social environment within the club is governed by contractual rules. This should reduce many of the risks to quality of life and property investment borne by residents. The club is designed according to the specification of developers, landowners, investors and any regulating government agency and if its benefits are accurately pitched and costed it acquires members. The design specification may subsequently be modified by residents within the constraints of the initial constitution.

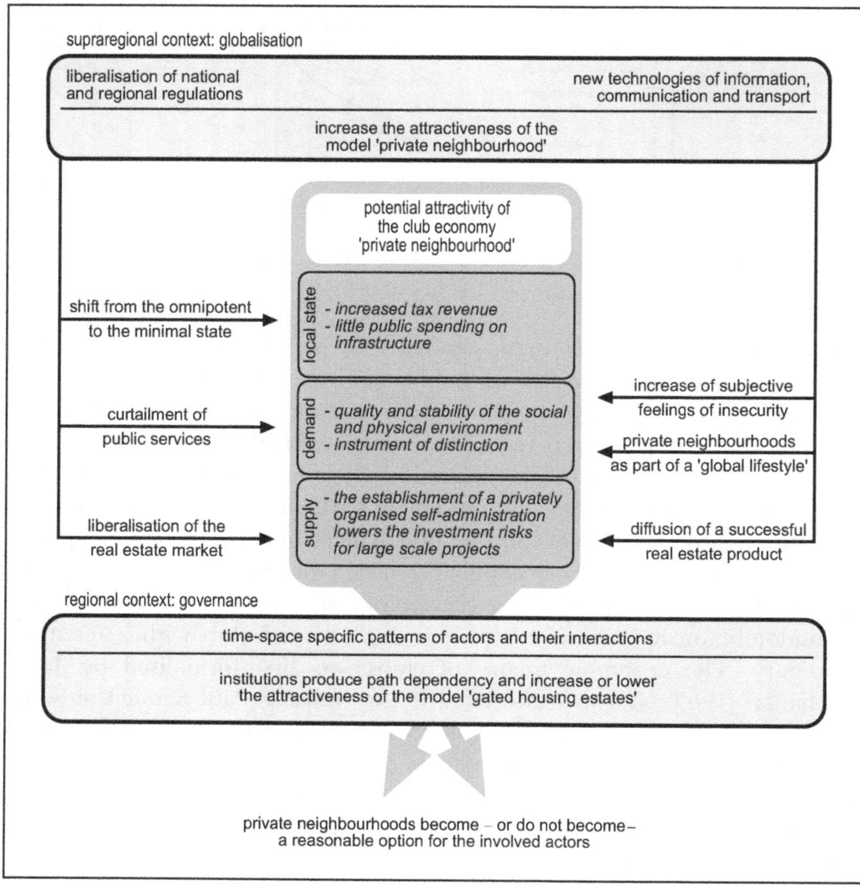

Figure 15.2 The attractiveness of clubbing together in cities. Source: Glasze 2003b

Where there are many club communities in a city and a competitive market develops, the price of the various packages of club goods and services may be expected to reflect the economic cost of supplying them and both choice and economy should be improved as a result.

As well as being attractive to home-buyers and renters and to the firms who develop, sell and manage private neighbourhoods, the residential club as an institution also offers advantages to the local state (figure 15.2). It frees the state from the obligation of supplying civic goods and services in certain localities. It also transfers some of the costs of local land-use control and other environmental and social regulations to the private sector.

The rapid development of gated housing estates in many regions of the world suggests that there are globally diffused and possibly global-scale processes at work. And yet the local variety captured by the chapters in this book and in our other work (see for example the collection of papers in *Environment and*

Planning B, 29 (3), 2002) clearly also suggest strong local influences. We find it helpful to understand these processes by focusing on the emergence of the club as a new device for organising interaction in cities (see Webster 2003; Webster and Lai 2004: chapters 5 and 6 for a theoretical discussion). What shift in conditions made it possible for the residential club to become an urban governance institution with mass appeal in many regions of the world? Three major factors can be pointed to (figure 15.2; Glasze 2003a).

First is the shift towards a minimal state and the associated trend towards decentralised governance. During the second half of the 20th century a model of the state developed under the influence of the political economic doctrine of Keynesianism. The welfare state rapidly spread into municipal affairs and the security function of government was extended to ensuring the security of minimum standards and more generally of 'opportunity'. Provoked by a crisis of finance and legitimacy, and by an ideological shift towards a pro-market doctrine, the interventionist state retreated, however. The breakdown of the communist systems in Europe weakened intellectual and political resistance to this retreat, and deregulation, privatisation and decentralisation became dominant trends in the evolution of governance institutions in countries the world over. All this may be understood as a process of experimentation with the mechanisms by which society orders its affairs (following the seminal ideas of institutional historians such as Douglas North – North and Thomas 1973; North 1990, 1992 – and earlier ideas of the Austrian and Chicago economists). The experimentation was not coordinated or planned, nor has it been systematic or comprehensive. It is made up of local initiative and invention; of local political solutions and popular movements; of trial and error in the making and changing of policy and laws; and of the actions of firms, individuals, governments and government agencies searching for the means to improve themselves and those whose needs they seek to meet. The process of rediscovering or refining what is understood to be good government is global. If one wanted to suggest a universal reason for it, the idea that societies will tend in the long run to look for forms of organisation and governance that tend to reduce the costs of social cooperation, including resource transaction, is a worthwhile direction to explore (Webster and Lai 2004: chapters 1–4; Webster 2003). Short of this, however, it may simply be observed that models of governance have constantly evolved throughout the history of civilisation and that the dominant search in contemporary societies of all political hues is for an acceptable boundary between state and market.

The growth of a private-neighbourhood industry may be understood in this context. In some countries, the liberalisation of real-estate regulation and urban planning has enlarged the margin of manoeuvre for private investors and permitted experimentation with an exotic idea. The attraction of a European and, more lately, a US lifestyle seems to have been a strong influence on historical and contemporary gated developments in Latin America (see the Latin America chapter in this volume). In transitional economies such as eastern Europe (see the chapter on Russia) and China, private neighbourhoods are variously an exotic implant, an adaptation of existing institutions and an experiment in market-sensitive urban governance.

In many regions, especially in the developing world, private residential complexes have become important substitutes for the local state, providing civic goods and services including regulation and security. Many studies have shown that the secured provision of high-quality infrastructural services (electrical power, drinking water, communication, local environment and education) plays a crucial role in making private and gated housing estates attractive to home-buyers (see for example Janoschka 2002; Leisch 2002; Glasze 2003a; and chapters in this volume). For many in the developing world's burgeoning professional and middle classes, buying into a private neighbourhood is the surest way of ensuring the kind of local amenities that their counterparts in industrial nations have come to expect from public government.

The second reason why residential clubs have assumed growing importance as an urban governance institution is the growing feeling of insecurity in cities throughout the world. In the course of growing social differentiation and individualisation grow fundamental uncertainties. Informal social networks, like kinship or other traditional forms of community, are getting weaker and are no longer available with certainty in times of crisis. At the same time, in many countries, formal systems of social security are also weakening. Uncertainties and fear are compounded by a competitive media industry that leads to a growing 'scandalisation' of media coverage. Adding to these factors is the global flux of migration, which increases the visibility of strangers and rouses new feelings of uncertainty (Siebel 2003). Against this background many inhabitants of guarded housing estates seem to be looking for security – the security of living in an environment in which the physical and social qualities are regulated by private contracts and perceived to be more stable than in 'normal' neighbourhoods.

In this context, it is not surprising that residential clubs have flourished. They are in many ways an institution for our time. They are solidly a capitalist institution in that they exist to assign property rights more clearly. In so doing they apparently reduce some of the perceived risks of living in open neighbourhoods – risks to continued enjoyment of certain levels of security and well-being. Their ingenuity is that they provide a mechanism for assigning rights over collective goods without compromising the institution of private property. Being part of a 'smaller public' helps residents feel more secure. Many studies report that a sense of community is a consideration in moving into a CID, though not necessarily a strong one. Having said this, it is interesting to note that contrary to much of the media coverage about secured communities, security seems to be only one motivation for moving into this form of neighbourhood. It is often not as important as the desire to secure the supply, more generally, of a bundle of rights including the rights to goods and services and freedom from risk of all kinds, such as the risks associated with neighbours who have different preferences and behaviour.

A third reason for the success of residential clubs is that the institution has become successfully commodified, marketed and subsumed into global and local cultures. Guarded housing estates are part of a global culture. Throughout the developing world they are marketed as places of modernity and of Western cultural values. Giroir's chapter in this volume captures this particularly well. In industrialised countries they are marketed with cultural references to tradition (as in neo-historical complexes or in schemes constructed around historical buildings), to modernity (as in chic central city schemes) or to particular lifestyles (like the 'golf-communities' or 'active-adults communities' in the USA and Canada). Usually there are references to elite or prestige values and these vary with local culture. The adverts in newspapers in Shanghai, Beirut and New Orleans would attract little interest if used to sell a scheme in London or Frankfurt but the adverts in the UK and German press do much the same thing using different cultural references. The role of the real-estate industry in spreading the culture of club-living is viewed critically by many commentators, including some of the authors in this book. Any such analysis should distinguish between good and bad practice and between market failure and firm or product failure, however.

Bearing in mind the general retreat of the state, and public governments' inability to supply adequate levels and mixes of civic goods and services, including security, entrepreneurs have naturally been active in looking for business in the voids. We characterised externalities above as missing markets. With appropriate technological and institutional invention, missing markets can be turned into markets – as with toll roads and quota trading. This happens when mechanisms are devised for controlling access and collecting payment. Externalities are business opportunities waiting to happen. The development, finance, marketing and management firms that created private-neighbourhood markets have made a business out of local public goods and externalities – positive and negative. By territorially enclosing and legally constituting a group of neighbours into a contractual community, revenue can be made from neighbourhood conflict and from neighbourhood agglomeration benefits. Regulations and communal services can be supplied according to price and preference. In designing their wares, residential-club entrepreneurs obviously need to pay close attention to the kinds of regulation, goods and services that their consumers want. This leads to niching and specialisation and to product packaging and standardisation.

The economic rationale for packaging and standardisation needs to be understood accurately – rather than being summarily dismissed for normative reasons (in the same way that burger chains tend to be). Packaged neighbourhoods are successful for similar reasons to packaged holidays. Neighbourhoods, like holidays, are what economists term 'experience goods'. Their value to a consumer can only really be gauged through the experience of consumption. This is not so with simpler goods with more singular qualities such as fuel and chocolate. Private-neighbourhood suppliers have discovered that packaging a bundle of neighbourhood 'products' (goods and services) and giving the bundle a distinctive theme and brand name saves buyers costs incurred when searching for a place to live or invest in and saves the even greater costs of experimentation. The chances of a satisfactory holiday experience for many are increased when selecting a clearly branded and contractually secured bundle of holiday attributes (location, style of hotel, view from room, on-site facilities, car hire and local excursions) from the brochure of a respected firm. The package holiday industry has flourished as entrepreneurs have discovered ways of giving consumers value for money and choice of a selection of holiday purchases in a way that lowers the risk of loss.

Economy of consumer search costs is a strong explanation for the rise of industries which bundle, package and brand complex multi-attribute commodities. This should be distinguished from the discussion of bad practices. McKenzie's work reveals many examples of the latter, including impenetrably long, complex and sometimes misleading contracts; bullying and unreasonableness in the exercise of HOA governance powers; poorly designed constitutions; and unreasonable municipal government pressure on communities to gate and incorporate. The manner in which a holiday is designed, marketed, sold and organised may have faults, particularly in the early days of the package tour industry. As more operators get involved, however, the industry becomes more competitive and this tends to weed out poor products and processes and bad practices. The housing market is certainly more constrained than the market for tour operators – choice is highly dependent on financial, cultural and social capital; the transaction costs are very high so that 'wrong decisions' are not easy to change; and, in the end, housing is a special good in so far as nobody really can abstain from consuming it. Nevertheless, there is competition between developers offering private neighbourhoods so that one might expect to some degree comparable processes in the housing market as in the tour operator market. It may not, however, and this is an interesting line of empirical enquiry. In an industry with strong scale economies, monopolistic practices can inflate price above marginal cost and restrict consumer choice and quality. This is a latent danger in the packaged neighbourhood industry for obvious reasons and may ironically be more of a problem where the scale of the industry in a city is constrained by regulation.

Packaging and advertising can also conceal the hidden health hazards in a product that has otherwise compelling immediate attractions. Exotic beach holidays entice consumers to overdose on ultraviolet radiation and burger chains encourage the consumption of harmful amounts of unsaturated fats. But herein lays another illustration of how society adapts in unpredictable ways. Government and private campaigns have been important in shifting the cultural attitudes towards smoking, fast food and sunbathing following public discussion of the evidence.

Concluding comments

The chapters in this volume point to a range of concerns about the use of clubs as institutions of urban governance. The volume also demonstrates why they are so appealing to so many. Having read hundreds of papers addressing these issues, visited gated communities the world over and interviewed expert academics, residents, government officials and industry professionals, we take the view that the focus of debate and investigation should be not only the immediate processes of urban change associated with gated developments but also the wider adaptation of cities. There is a subtle but important distinction. Much analysis of urban change is comparatively static and fails to consider changes happening in other parts of the system. Regulation might have counter-intuitive outcomes and is expensive. Research on private neighbourhoods needs to help build the evidence base for enlightened public discussion. This will require well-formed theories of markets and governments, households and firms, institutions and organisations. The theory needs to be explicitly spatial since the private and social costs and benefits are sensitive to distance. It also needs to be explicitly dynamic, understanding that organisational, institutional and spatial order co-vary in complex ways over time, as society searches for cost-effective ways of organising itself in pursuit of shared, but contested, goals. The idea that many different urban societies and many different income groups within them are choosing the ultimate form of decentralised urban governance is an intriguing one. Among the effects are the redrawing of boundaries of contestation, a new spatial and organisational tier of collective action and transaction, a reduction in conflict over shared resources as property rights become more clearly defined and, at the same time, an intensification of conflict as disputes arise at the boundaries of these more finely grained entitlements. Something like entropy seems to be at work and its dynamic has a distinctly fractal dimension. Residential clubs have the effect of establishing property rights over public-domain urban resources and in so doing they reduce dissipative competition but at a price. More closely defined ownership means more proprietary boundaries, however, and boundaries are inevitably the scenes of dispute. Thus a society with deepening institutions and deepening proprietary order spawns deepening conflict – for the contracts that define boundaries of ownership can never be complete. The growth path can probably be described mathematically by a power function. The potential for conflict rises according to some power of the number of subdivided ownership rights. In a grid-square pattern of urban land parcels, the number of ownership boundaries in a piece of land subdivided into four separately owned plots, for example, increases from four to sixteen. Each new boundary is a site of potential public-domain dispute (Webster and Lai 2004: chapter 7). The increasing cost of establishing social order is entropic. Increasing degrees of order as a result of institutional evolution is not of course entropic – quite the opposite. But if that order is only achieved at an expense and if that cost rises at a rate greater than the welfare it secures, then in the long run the process is dissipative. This is the curse of the city: benefits of agglomeration at the expense of crowding and congestion. At the extreme, unrestrained competition (for road space, housing, clear air and water and so on) can completely dissipate the benefits. Only investment can avoid this – investment in infrastructure and in the institutions of market and government that produce greater order. Private neighbourhoods are part of this evolutionary story. They resolve conflict but they generate different forms of conflict. They raise the predictable costs of urban living and raise the benefits commensurately but create unpredictable costs and benefits. They are not unique in this at all; the same may be said of all inventions that try to reverse the entropy of the city. Some may be more stable and more cost-effective than others.

Bibliography

Arrow, K.J. (1951) *Social Choice and Individual Values*, New York: John Wiley.

—— (1970) 'The organisation of economic activity: issues pertinent to the choice of market versus non-market allocation', in R.H. Haverman and J. Margolis (eds) *Public Expenditure and Policy Analysis*, Chicago, IL: Rand McNally: 59–73.

Briffault, R. (1999) 'A government for our time? Business improvement districts and urban governance', *Columbia Law Review*, 99 (2): 365–477.

Buchanan, J. (1965) 'An economic theory of clubs', *Economica*, 32: 1–14.

Chen, C.Y. and Webster, C. (2005) 'Homeowners associations, collective action and rent seeking', *Housing Studies*, 20 (2): 205–20.

Dear, M.J. and Flusty, S. (1998) 'Postmodern urbanism', *Annals of the Asscociation of American Geographers*, 88 (1): 50–72.

Deng, F.F. (2002) 'Ground lease-based land use systems versus common interest development', *Land Economics*, 78: 190–206.

—— (2003a) 'The rebound of private zoning: Property rights and local governance in urban land use', *Environment and Planning A*, 35: 133–49.

—— (2003b) 'Collective goods and the political hold-up problem', *Journal of Institutional and Theoretical Economics*, 159 (2): 414–34.

Frantz, K. (2000) 'Gated communities in the USA: A new trend in urban development', *Espace, populations, sociétés*, 1: 101–13.

Friedman, John and Goetz, Wolf (1982) 'World city formation: An agenda for research and action', *International Journal of Urban and Regional Research* 6: 309–44.

Glasze, G. (2003a) 'Private neighbourhoods as club economies and shareholder democracies', *BelGeo*, 1: 87–98.

—— (2003b) *Die fragmentierte Stadt. Ursachen und Folgen bewachter Wohnkomplexe im Libanon*, Opladen: Leske and Budrich (Stadtforschung aktuell 89).

Glasze, G., Pütz, R. and Rolfes, M. (eds) (2005) *Stadt – (Un-)Sicherheit – Diskurs*, Bielefeld: Transcript.

Hamnett, C.R. (1994) 'Social polarisation in global cities: Theory and evidence', *Urban Studies*, 31: 401–24.

Janoschka, Michael (2002) *Wohlstand hinter Mauern: Private Urbanisierungen in Buenos Aires*, Wien: ISR (Forschungsberichte 28).

Kanitschneider, S. (2002) '*Condominios* und *faccionamientos cerrados* in Mexiko-Stadt – Sozialräumliche Segregation am Beispiel abgesperrter Wohnviertel', *Geographica Helvetica*, 57 (4): 253–63.

Kohler, P. (2002) 'Geschlossene Wohnkomplexe in Quito – Naturraum und rechtliche Rahmenbedingungen als Einflussgrößen für Verbreitung und Typisierung', *Geographica Helvetica*, 57 (4): 278–89.

Leisch, Harald (2002) 'Gated communities in Indonesia', *Cities* 19 (5): 341–50.

McKenzie, E. (2003) 'Common interest housing in the communities of tomorrow', *Housing Policy Debate*, 14 (1–2): 203–34.

Marcuse, Peter (1997): 'The ghetto of exclusion and the fortified enclave: New patterns in the United States', *American Behavioral Scientist*, 41 (3): 311–26.

North, D.C. (1990) *Institutions, Institutional Change and Economic Performance*, Cambridge and New York: Cambridge University Press.

—— (1992) *Transaction Costs, Institutions and Economic Performance*, San Francisco, CA: International Center for Economic Growth.

North, D.C. and Thomas, R.P. (1973) *The Rise of the Western World: A New Economic History*, Cambridge: Cambridge University Press.

Postrel, V. (1999) 'After socialism', *Reason*, 31 (6): 32; cited in B. Caldwell (2004) *Hayek's Challenge: An Intellectual Biography of F.A. Hayek*, Chicago, IL: University of Chicago Press: 368.

Salcedo, R. and Torres, A. (2004) 'Gated communities in Santiago: Wall or frontier?', *International Journal of Urban and Regional Research*, 28 (1) (March): 27–44.

Samuelson, P.A. (1954) 'The pure theory of public expenditure', *Review of Economics and Statistics*, 36: 387–9.

—— (1955) 'Diagrammatic exposition of a pure theory of public expenditure', *Review of Economics and Statistics*, 37: 350–5.

Siebel, W. (2003) 'Die überwachte Stadt – Ende des öffentlichen Raums', *Die alte Stadt*, 30 (3): 247–57.

Soja, E. (2000) *Postmetropolis: Critical Studies of Cities and Regions*, Oxford and Malden: Blackwell.

Tiebout, Charles M. (1956) 'A pure theory of local expenditures', *Journal of Political Economy*, 64: 416–24.

Walters, M. and Kent, P. (2000) 'Institution economics and property strata title – a survey and case study', *Journal of Property Research*, 17 (3): 221–40.

Webster, C. (2003) 'The nature of the neighbourhood', *Urban Studies*, 40 (13): 2591–612.

Webster, C. and Lai, L.W.C. (2004) *Property Rights, Planning and Markets: Managing Spontaneous Cities*, Cheltenham and Northampton: Edward Elgar.

Yap, K.S. and Sakchai, K. (2000) 'Bangkok's housing boom and the financial crisis in Thailand: Once only the sky was the limit', *Housing Studies*, 15 (1): 11–27.

INDEX

For Product Safety Concerns and Information please contact our EU representative GPSR@taylorandfrancis.com Taylor & Francis Verlag GmbH, Kaufingerstraße 24, 80331 München, Germany